Tropical Agriculture

Geographies for Advanced Study
Edited by Professor Stanley H. Beaver, M.A., F.R.G.S.

The British Isles: A Geographic and Economic Survey
Central Europe
Concepts in Climatology
East Africa
Eastern Europe
Geography of Population
Geomorphology
The Glaciations of Wales and Adjoining Regions
An Historical Geography of South Africa
An Historical Geography of Western Europe before 1800
Human Geography
Introduction to Climatic Geomorphology
Land, People and Economy in Malaya
The Landforms of the Humid Tropics, Forests and Savannas
Malaya, Indonesia, Borneo and the Philippines
North America
The Polar World
A Regional Geography of Western Europe
The Scandinavian World
The Soviet Union
Statistical Methods and the Geographer
Structural Geomorphology
Tropical Agriculture
The Tropical World
Urban Essays: Studies in the Geography of Wales
Urban Geography
West Africa
The Western Mediterranean World

Tropical Agriculture

A geographical introduction and appraisal

Walther Manshard

Professor of Geography and Director of the
Department of Geography of the University of Freiburg

(formerly Director of the Tropical Institute of
the Justus Liebig-University of Giessen and of
the Department of Environmental Sciences, UNESCO, Paris)

Translated by Mrs D. A. M. Naylon

Longman

Longman
1724-1974

LONGMAN GROUP LIMITED
London
and LONGMAN INC., New York
Associated companies, branches and representatives
throughout the world

© Bibliographisches Institut AG, Mannheim 1968

Translation © Longman Group Limited 1974

First published 1968 by Bibliographisches Institut AG, Mannheim
English edition first published 1974

ISBN 0 582 48187 2
Library of Congress Catalog Card Number: 73–88902

Phototypeset by Filmtype Services Limited, Scarborough, Yorkshire
and printed in Great Britain
by Lowe & Brydon

Contents

List of figures

List of plates

Editor's Note

The German version of this book was published in 1968, and dedicated to one who may perhaps be regarded as the founder of the study of tropical agricultural geography, Leo Waibel (1888–1951). The lapse of time between the German and English publications has enabled Prof. Manshard, profiting by the wide travels that his UNESCO post entailed, to revise the work appreciably and incorporate the results of his own further researches and those of other workers. The adoption of a larger format than that of the German original, and the necessity for redrawing after translating the inscriptions, have enabled many of the maps and diagrams to be much improved. The extensive bibliography contains a great deal of German work, a lot of which may not be readily accessible to the English reader; but at least it will help the specialist, and it gives some impression of the immense contribution to tropical geography that German scientists have made. The book will form an excellent companion to Gourou's *Tropical World* in this series.

S. H. BEAVER

Foreword

Any traveller who flies nowadays from Europe to the tropics in a matter of a few hours will never forget the striking scenic differences in the natural and cultural landscapes he passes over. Leaving behind the dry, empty deserts, he passes over extensive savannas where the farming population has developed land use systems and farm types which vary greatly from area to area; finally he reaches the dense evergreen rainforests near the equator. The distribution of land and sea, relief, climate and vegetation, and—last but not least—man's impact, characterise these tropical environments.

A comparative survey of this areal structure is one of the main tasks of geography, and the analysis of both natural and socio-economic features represents an important contribution to the preparation of development aid. The great number of studies dealing with the individual symptoms and problems of tropical agriculture render such a synopsis very difficult. In this book an attempt has been made to provide an introduction which will help in the many aspects of investigation into the agricultural geography of the tropics.

The first stimulus to write this study was provided by lectures on 'The comparative agricultural and social geography of the tropics', given at the Universities of Accra, Liverpool, Cologne and Giessen.

As there are very few comprehensive agricultural geographical studies of the tropics in international literature, the attempt had to be made to find a form for this which would stimulate both the professional geographer and the student of geography, the teacher and the reader who is interested in issues concerning the developing countries. Therefore in some parts the narrower limits of agricultural geography have been transgressed. The general methodological and historical introductions to agricultural geography have been kept as short as possible.

The author's interest in tropical Africa is reflected in the list of contents, an interest which has been strengthened by eight years' service in West Africa. In presenting examples of various phenomena, reference could be made to previous studies by the same author; but as the object was a well-balanced geographical picture of tropical agriculture, needless to say, reference had to be made to the work of many other authors.

This is not intended as a general text for students, introducing techniques and concepts or applying the more general economic principles in the field of agricultural geography. It rather attempts to present a synopsis of tropical agriculture by examining the social, economic and environmental factors important to the geographer. It aims at giving specific regional examples rather than general models. With the lack of sufficient statistical information and the scarcity of comparative studies for many tropical countries the more empirical method of geographical synthesis has been used to present land use patterns and farming systems in their spatial settings and historical contexts, also with a view to evaluating their possibilities for agricultural development.

When this book first appeared, in 1968 in German under the title *Einführung in die Agrargeographie der Tropen*, nearly all reviews of it in the international journals were favourable because at the time there were very few texts of a comprehensive character in existence. As several reviewers suggested a translation into English, Professor Beaver, editor of the Longman's series of Geographies for Advanced Study, took the initiative to arrange for this. For a number of personal reasons, the translation was considerably delayed. Because of his appointment with Unesco, and being away from the 'home base' of his university, the author—much as he would have liked to do so—did not have the time to rewrite the book. However, several parts, including the chapter on the 'Green Revolution' and other regional examples have been added. Certain passages intended principally for the German reader have been omitted or summarised. Also in the past few years many new studies on tropical agriculture have appeared, some of which are included in the bibliography.

I should like to express my thanks to numerous colleagues and collaborators, my assistants and older students for discussions held with them and for their criticisms, which were appreciated; I would also like to thank Frau Herberich and Messrs Desselberger, Grenzebach, Küchler, Mäckel, Meyer, Möller, Schliephake, Schmitz and Weyl. Last but by no means least, I should like to express my gratitude to Professor Beaver as editor, and to Mrs Dorothea Naylon who mastered the difficult task of translating the book.

WALTHER MANSHARD

1
The tasks and aims of agricultural geography in the tropics

'Agricultural geography is the science of that part of the earth's surface formed by agriculture, seen both as a whole and in its parts, in its outward appearance, in its inner structure and in its intricacies' (Otremba, 1960, p. 25).

Being one of the component disciplines of economic geography, agricultural geography combines elements and methods which are found in geography, economics, and agricultural science. One of the chief objectives of agricultural geography is to examine spatial differences in the various manifestations of agriculture. In this connection, however, the term 'agriculture', meaning one of the essential appearances of the earth's surface, is used in a wider sense than is the case in pure agricultural science. Originally mainly based on the investigations of natural geographical factors, then making regional statistical enquiries into the areas cultivated for plants of economic value or the ratio of population to animals, agricultural geography gradually began to adopt ecological methods of research, examining, for instance, the widely differing farming and land use systems of the world in their relationships with their environment.

Along with the development of geographical methods of investigation, the problems and views of agricultural geography have also undergone changes in recent decades. Research into the causal connections and reciprocal effects of nature, man and his economy were pushed more into the background. Going beyond the analysis of natural causal connections, research into geofactors, in the broadest sense of the word, has gained more importance in the methodological developments of recent years, as also has research into specific social and economic conditions. This is the direction taken by modern geography in its efforts to understand the spatial distribution of geographically relevant social structures and the spatial behaviour of social groups. For agricultural geographical research this question of economic behaviour has recently regained importance in discussions about 'development mentalities' in the context of the social economy of development countries (cf. E. Wirth, 1965a, b, and other works, e.g. *Wirtschaftsgeist*, Rühl, 1925). In particular, the polarity of progressive and traditional attitudes in neighbouring regions can often be traced back to different group attitudes or behaviours, which social-geographical categories often help to explain, for example, the contrast between developed and underdeveloped regions in areas of different religious practices, or the appearance and extensions of innovation in certain communities.

With the present emphasis on diversification and import substitution the possibilities of success for such policies depend much on the agricultural traditions and the historical and economic development of a given country or region. The agricultural geographer can do a lot to investigate these aspects, comparing, for instance, the different patterns of spatial, social and cultural organisation of plantations, peasant smallholdings or other forms of enterprise.

In recent years mathematical modelling has been extensively used in agricultural geography. This theoretical approach, using the techniques of statistics and descriptive mathematics, has become valuable as a research tool particularly in regions with good statistical documentation such as North America and parts of Europe. For the tropics these methods have so far been applied only rather reluctantly and often more for a theory for agricultural regionalisation. Because of the relatively historical and cultural orientation of geography in France and Germany, these new analytical techniques have not been used there to the same extent as by English writers (Gregor, 1970).

Because of the lack of enough reliable statistical data in the humid tropics, a quantitative approach is so far feasible only in exceptional cases. For the more comprehensive and comparative studies of the mid-1960s, such a presentation was not possible. For more recent attempts, the reader is referred to the books of Gregor (1970), Morgan and Munton (1971), and the more specialised literature.

The agricultural geography of the tropics is relatively young as a branch of general geography. The classic works of the geographers Alexander von Humboldt and Carl Ritter contain a great deal of information and comment on this field of study. Important contributions have also been made in sister disciplines, especially cultural history and economic history (such as Viktor Hehn, 1813–90, and Eduard Hahn, 1856–1928) and among the great agricultural theoreticians (J. H. von Thünen, 1783–1850). But agricultural geography in the proper sense of the word (and including the tropics) has only existed for the last few decades. Besides important pioneers such as T. H. Engelbrecht and the American O. E. Baker, who were the first to introduce absolute and comparative methods of cartographic representation into agricultural geography, we must first call to mind Leo Waibel (1888–1951) who, together with his followers, can be regarded as the founder of modern agricultural geography in three continents, and whose research on tropical Africa and tropical America was of international repute and had many followers.

The question of whether there really is a 'geography of the tropics', or indeed whether it can exist at all, has been discussed from time to time. Gourou states in his classic work *Les pays tropicaux: principes d'une géographie humaine et économique* (English translation *The Tropical World*), that the tropics have their own physical and cultural geography. Nevertheless, sufficient attention has not yet been paid to research into specific characteristics of the tropics. Indeed, little research has been done so far in the comparative geography of tropical agriculture, and thus it is not possible to draw up a refined typology comparable to European conditions. In spite of being uniform in their natural features, the extensive tropical climatic and vegetation belts show a distinct differentiation according to cultural and social geographical phenomena which are only indirectly related to the physical and biogeographical conditions. Many human communities with different ways of life and at different stages of development orientate themselves in their surroundings in different ways from us. Some tribes in West Africa, for instance see the 'geography' of their country, above all, in terms of social structure (Bohannan, 1954, 1963; Manshard, 1965a). Their 'map' is a kind of genealogical map of different productive areas and parcels, whose boundaries can vary from year to year according to social changes.

The question thus arises of how far tropical areas with similar natural conditions can be compared. As can be expected with different national, social and historical conditions, the elements which predominate in the cultural patterns are those which, like land use, for instance, are closely related to natural conditions; but even where social and historical factors play a major role, astonishing analogies or parallels can be discovered.

It is important, in fact, to compare different tropical areas in the world precisely from these points of view in order to demonstrate both parallels and differences, through the expansion of certain typical characteristics, and to explain the special structure of isolated cases by looking at them in the context of a more extensive area.

2
The geographical limits of the tropics and subtropics

The conventional solar mathematic definition of the tropics as the area between the meridians of Cancer and Capricorn was established in the ancient world by Parmenides and Aristotle, on the basis of astronomical observations. L. v. Buch (1829) was the first to connect the tropics more closely with high precipitation, introducing at the same time the term 'subtropics'. It was many years later before Supan (1879) defined the tropics by means of more exact climatological criteria. The boundary suggested by him follows the 20° isotherm of average annual temperature, which often coincides with the polar limit of palm trees, and he thus includes within the tropics a much larger area than usual. Köppen (1931), on the contrary, defines tropical rainy climates (which he calls A, in contrast to dry climates, which he calls B) as those areas of the earth's surface with an average temperature in the coolest month of over 18°C, plus ample precipitation. Köppen distinguishes two tropical rainy climates: on the one hand the humid and hot climate (Af-) of the tropical rainforest, where although there are seasonal variations in temperature, precipitation is at least 60 mm even in the driest month; and on the other hand the periodically dry tropical climate of the savanna (Aw), in which the symbol 'w' signifies a distinct dry period in winter (at least one month with less than 60 mm rainfall).

Since Köppen numerous further attempts have been made to define the tropical zone more exactly. One such effort at delimiting the tropics is that of Philippson (1933), using statistical climatological methods. He indicates the 24° isotherm of the coolest month as the limit of the inner tropics, and the 20° isotherm as the limit of the outer tropics. Other authors (Flohn, 1957) define the tropics and subtropics in terms of dynamic meteorological climatology, using predominant winds (already suggested by Hettner, 1930, 1934) and air masses.

The classifications of Paffen-Troll (1963, 1965) and Thornthwaite (1948), despite their different basic conceptions, use a combination of moisture and temperature characteristics. Troll (1943) makes up for some deficiencies in the ideas of Köppen whose climatology, based on averages, neglects the vertical differentiation of climate in the tropics; he proves that at all altitudes in the tropical zone the daytime climate (gross daily tempera-

4

ture amplitude, in contrast to a smaller yearly amplitude) is constant. If one compares the mean yearly and daily fluctuations of many stations, lines of 'balance' can be drawn in the northern and southern hemispheres on which all places show the same daily and yearly fluctuations in temperature. Troll has used this boundary line of predominant daily/annual fluctuations in temperature in his delimitation of the tropics, locating it by interpolating thermal isopleth diagrams of different stations (Troll, 1943). These diagrams are obtained, in turn, by combining time of day and month with their respective temperatures in a system of coordinates.

Lauer (1952) and Garnier (1961a, b) attempt a more subtle subdivison of parts of the tropics. Lauer distinguishes between humid ($N > V$) and dry months with the help of Martonne's aridity index ($12n/(t+10) = 20$).[1] The classification is then done by lines connecting places with the same number of humid months (isohygroms). Garnier improves on this method by applying Thornthwaite's calculations combined with his own method of estimating potential evapotranspiration.

It is relatively easy to define the tropics as the zone which does not have thermal seasons (the thermal tropics), although it must be borne in mind that oceanic influences also weaken the contrasts between thermal seasons. It is very difficult, however, to delimit the tropics by means of hygric (moisture) seasons. Even if it were possible to state the humidity exactly (for instance, by following Garnier's method), some thermally nontropical regions would have to be counted as tropical because of their humidity; and similarly, some thermally tropical and subtropical regions would be shown as belonging to the dry zone because of their aridity.

When the Unesco commission, under Meigs and Amiran, examined the 'arid zone' of the outer tropics, similar problems of delimitation were found.[2] Above all, the problem arose of whether the desert belt of the earth should be included at all in the tropical zone proper. The fluctuating positions of the various boundaries of the dry zone raise yet another question. It is true that cultivation as practised in rainy climates can penetrate, with

---

[1] n = mean monthly precipitation, t = mean monthly temperature.

[2] The total area of the arid tropics can be variously stated, depending on the definitions applied. It occupies between 26 per cent (Trewartha, 1943) and 36 per cent (Shantz, 1956) of the surface of the continents (excluding Antarctica). Shantz classifies the arid tropics:

(a) According to climate:		
1. Semi-arid	21	million sq km
2. Arid	22	million sq km
3. Extremely arid	6	million sq km
Total	49	million sq km

(b) According to vegetation:		
1. Semi-arid (thornbush, short grass)	7	million sq km
2. Arid (desert grass)	33	million sq km
3. Extremely arid (desert)	6·5	million sq km
Total	46·5	million sq km

the aid of special methods (dry-farming), into areas with only 200–300 mm annual precipitation; often the zones of real or viable agricultural land use are very much smaller.

It is easy to see that there are as many tropical boundaries as there are tropical characteristics. To attempt a further subdivision of the whole tropics leads to a similar conclusion. This task has been carried out more intensively in recent years (see map, Fig. 2.1).[1] Garnier (1961b) attempts a climatic delimitation of the humid tropics. He divides them into two primary zones, according to the number of wet-tropical months, and three secondary zones, according to the number of rainy months (over 75 mm per month) and the mean annual precipitation. Later he defines a wet-tropical month as the most important time unit, based on the following criteria: (1) mean monthly temperature of over 20°C, (2) mean relative humidity of over 65 per cent, (3) mean vapour pressure of over 20 mb. If all twelve months of the year show these conditions, there is no doubt that the place in question lies in the humid tropics. It still remains to be clarified how long a period of time is sufficient to ascribe a place to the humid tropics, if it has a humid tropical climate for less than twelve months. According to the findings of Garnier and Budyko (1958), for a place to be regarded as belonging to the humid tropics it has to fulfil the first of the above conditions for at least eight months; there should be a mean vapour pressure of > 20 mb and a mean relative air humidity of over 65 per cent for at least six months. At the same time the mean annual precipitation in the humid tropics should be at least 1 000 mm. In particular, essential ecological factors for the tropics are the amount and intensity of precipitation in the form of rain.

Holdridge (1959) has worked out a system of classifying tropical habitats. He starts from the physiognomy of the natural vegetation (the shape, grouping and associations of plants), wherein a number of variable factors reflect climatic conditions. Holdridge stresses the dependence of ecosystems on climate. In diagram form he shows the 'natural life zones' of different geographical latitudes and altitudes, with their dependence on temperature, precipitation and evapotranspiration potential. This model was developed further by Tosi (1964; see also Baker, 1966). More recently Jätzold (1970) has made a valuable contribution to a decimal classification of tropical agricultural climates using the climatic limits of coffee-growing in East Africa as an example.

Even more difficult than delimiting the tropics on the basis of more or less exact climatic values or maps of plant geography and physiognomy, is the definition of the term 'tropics' by economic and cultural criteria. Waibel (1937, p. 19) defined the tropics as those regions of the earth 'in which we find mature, economically valuable plants which require a lot

---

[1] Much research into a better comprehension and coverage of the humid tropics has had special support from the International Geographical Union and Unesco.

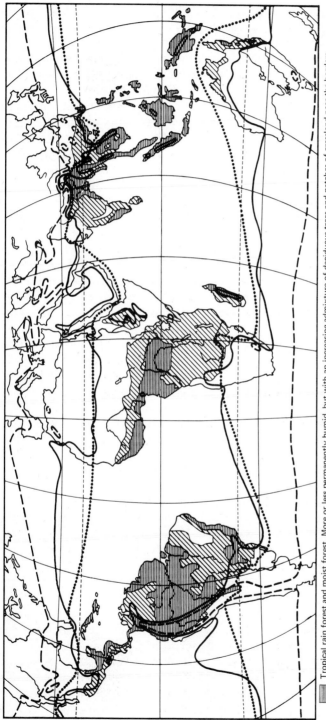

FIG. 2.1. Delimitation of the tropics and subtropics. By W. Manshard after Waibel (1937), Küchler (1961) and von Wissmann (1966)

of heat'. This European-influenced definition creates several difficulties. We know considerably more about the actual rather than the potential distribution of these valuable plants. In representing the areas where they are cultivated, formal floral boundaries seem to be stressed. However, many tropical annuals expand far beyond the limits of the tropics, finding conditions which, even if shortlived, are sufficient for their vegetative period. Areas in which we find mature tropical plants are of limited value in delimiting the tropics, since such plants never fully achieve their potential distribution. Moreover, their distribution can be changed constantly, either by the growing of new species or by man's influence on the natural environment.

Another difficulty is that statistical surveys of the cultivation of tropical plants are usually made with relation to political administrative units. The comparatively limited validity of Waibel's economic geographical definition of the tropics is illustrated by the boundaries shown in Fig. 2.1, which are the result of using the export figures of the countries concerned, and thus follow the political frontiers of the states. Otremba (1950) laid down a similar outer boundary for the tropics in classifying countries according to their economic geography. In this classification, Mexico and India fall entirely within the tropics, whereas southern China and northern Australia are excluded.

By way of comparison, a further delineation of the tropics and subtropics has been given in Fig. 2.1, according to climatic aspects. This follows von Wissmann (1948) who uses the absolute frost limit and the lack of heat limit (coldest month over 18·3°C) for demarcating the vegetational and agrogeographical warm tropics. As the boundary between the subtropics and the cool temperate zone (or mountain climates) von Wissmann (in Blüthgen, 1964, 1966) chooses a thermal criterion (eight months > 9·5°C). In representing the humid tropics I follow Küchler (1961), who separates constantly humid areas from periodically humid ones according to plant geography criteria.

A modern statistical delimitation of the humid tropics has also been proposed. Perhaps it will be possible one day to develop a model which contains numerous physical and economic geographical variables and which helps to define the term 'tropics' more exactly. The working out of the data could be done by computers.

Even more controversial than the delimitation of the tropics is the definition of the subtropics, climatologically as well as from the point of view of economic geography. As with the demarcation of the inner and outer tropics, for the subtropics, too, there are many possible boundaries, depending on the individual characteristics which are regarded as subtropical. Whereas Allisow (1954), for instance, who classifies climatic belts according to the distribution of air masses, would regard a large part of the USA, southern Europe and the Near East as part of the subtropics, separating them to south and north from the tropical and temperate air masses, other climatologists propose the complete exclusion of the desert belt from the tropical zone and its inclusion in the subtropics.

The zone of the subtropics is more difficult to delimit than the more uniform inner and outer tropics, because of its climatic characteristics as well as its much more differentiated economic and human geography. In oceanic regions and on the weather side of mountain ranges there are continuously humid subtropical areas with usually mild winters. Subtropical monsoon belts extend from south China as far as Korea and Japan. Subtropical regions with winter rainfall, also called Etesian regions after the north winds which blow in summer, characterise extensive areas of the warm temperate zone (e.g. the Mediterranean areas of Europe, southwest Australia, California, Central Chile, the South African Cape). Warm temperate areas with summer rainfall are often regarded as part of the subtropics (e.g. Gran Chaco, the Plate area proper, eastern Australia). The subtropics are not included in this study as they are often quite distinct in their physical character and economic life.

3
Outlines of agrarian development

In comparison with the development of mankind, which probably began in Africa over a million years ago and even in many tropical countries goes back for several centuries, agriculture proper is very young. In his earliest period of development man had no other weapons or equipment than simple tools of wood and stone, and the food he gathered was vegetable. In the course of time he evolved weapons for hunting, which enabled him to enlarge his diet and become a 'glutton for anything'. He usually lived near water, in caves, under overhanging rocks, in wind shelters of twigs and leaves, or in simple huts; and he learned the use of fire, which enabled him to live in colder areas hitherto denied to him. At this stage man probably penetrated for the first time into the earth's woodlands, where he used fire, perhaps unintentionally, to destroy the vegetation. Only at a much later and more specialised stage did he use fire purposely, first for hunting and then for shifting cultivation.

The game-hunting and food-gathering stages occupy an overwhelming proportion (98–99 per cent) of the economic development of mankind, and were still widespread in pre-Columbian times. Today, however, there are no more than a few tens of thousands of people living in this economic state, a number which has been rapidly decreasing due to the strong influence of modern science and civilisation. In a later chapter (p. 183) the aboriginal groups of Australian hunters and food-gatherers are described, to provide an example of this very important stage in the cultural and economic development of mankind. Even this description has become of almost historical interest, since nowadays many of these aborigines have to a large extent adopted modern economic methods.

Only since the Mesolithic, and more particularly since the last glacial period, has a growing adaptation of human ways of life to various climatic regions become apparent. In the temperate as well as in the tropical zones it led to a specialisation of fishing, hunting and food-gathering peoples. Prehistoric finds support the hypothesis that meat consumption rose rapidly among these people (Hötzel, 1963). The division of labour between men and women became more pronounced, as women devoted themselves more and more to gathering certain fruits and products. Increase in population frequently led to proper settlement and thus to a more sedentary way of life. When hunters and fishermen began to settle down, their protein-rich diet was partly replaced by a diet containing more carbohydrates of vegetable origin. Old hunting charms were replaced by fertility rites. After goats and sheep had been domesticated the stock of useful animals was enlarged by cattle, pigs and donkeys. Horse breeding definitely comes later.

An important 'germ cell' of the development of clan peasantry was the enlarged family, whose numerous household disposed of sufficient labour to allow a certain division of work. Planned cultivation and the growing of numerous wild plants safeguarded man's nutrition to an extent hitherto unknown. In the agrarian landscape the contrast between extensively and intensively used soils became more distinct. There is controversy about whether further development, from more advanced food-gathering to peasant cultivation with stock-rearing, took place through the cultivation of tropical plants apparently originating in southern Asia (India). Uncertain too is the origin of sedentary grain cultivation and the domestication of animals, for which there is evidence in the steppes of the Near East since the Neolithic. This development certainly did not take place uninterruptedly, but differed from region to region and may partly have happened in a series of jumps. Various cultivation techniques and agrarian ways of life were combined; an example is provided by the Red Indians of North America, who besides hunting, cultivated plants at the same time. In general, however, the production of crops started with sedentary groups. The semi-arid areas in particular, both in the tropics and in the warm temperate zone, are considered to be the nuclei of mankind's agrarian development, offering an adequate if not very wide range of food. An important exception is the growing of manioc and other starchy root crops in South America (probably in the Orinoco–Amazon region, according to Braidwood, 1962.

The part played by natural environmental factors in the agrarian development of mankind is an important question which has not yet been answered. We certainly notice in the early days of man's cultural development a distinct preference for the semi-arid and warm temperate climatic zones. New and important archaeological findings are now available from the New World (Central America) about the transition from food-gathering to sedentary cultivation (Braidwood, 1962).

In this connection, the very interesting question arises of why 'culture centres' such as those in southwest Asia or North America did not develop in other areas with similar natural conditions (for example, on the west coast of America or in northwest Africa). Were these areas too isolated? Did the intensive food-gathering economy offer too little challenge, as has been proved in the case of California? To what extent did the more intense cultural and economic development of some nuclear areas block that of others? In order to answer these questions many problems have to be solved and many blank spaces on the prehistoric archaeological maps of man's development have to be filled in.

Eduard Hahn's classic theory of related evolutionary economic stages, wherein animal husbandry followed the stages of hunting and food-gathering, is insufficient to explain many facts. For example, in some areas of the tropics and subtropics, especially in south and east Asia, nomadic pastoralism was a secondary development from already existing types of economy, and it never existed at all in the rainforests of the central tropics because of climatic conditions. Today it can be taken as proved that in the tropics and subtropics of the Old World nomadic pastoralism is to be regarded as an ecologically determined variant of peasant grain cultivation, which spread as a result of cultural exchange and migration and opened up drier landscapes to agricultural use. Animal husbandry did not develop from nomadism in some areas, for nomadism is often much more recent, appearing in different regions at very different times (Dittmer, 1965, p. 20). As a dominant type of economy, pastoral nomadism is found mainly in the arid and semi-arid areas of the Old World—in an area, that is to say, in which according to carbon 14 datings the first domesticated animals (pigs, dogs, chickens) were tamed 7 000 to 8 000 years ago. Southern or western Asia is thought to be the region of origin of cattle-breeding. Proven findings of domesticated cattle appear first in sedentary village cultures (ibid., p. 11). The zebu and water buffalo probably come from India, sheep and goats from southwest Asia, and the donkey from North Africa. The Arabian camel is derived from North African wild species.

In the New World one only finds llamas and alpacas in the Central Andean cultures, the former used mainly as pack animals, the latter reared for their wool. The characteristic animals of the North American prairies, the bison, remained undomesticated.

Right from the beginning there was tension between the sedentary groups and the pastoral nomads, who rode horses from the second millennium BC onwards and later used camels, and this tension often led to warfare. More frequently, however, under peaceful conditions they managed to supplement each other's economies. They found ways of living together, probably similar to the symbiosis between the herders and hoe-cultivators of East Africa. In the early days of pastoralism the unmounted herders were frequently dependent on sedentary tribes, or even derived from them. Later on, strongly organised warrior groups of nomads were uppermost, because of their greater power and mobility. Thus the nomadisation of extensive areas took place at the time of the nomadic conquests and the formation of states in Asia and Africa. However, the population of some urban centres (e.g. in the western Sudan) often defended themselves successfully against the attacks of the nomads. At certain times, especially when there was greater political stability and under the influence of certain ecological and economic factors, nomadic groups became sedentary again or adopted forms of partial or seasonal nomadism, with alternation of pastures (see p. 111). The Vedic Indian nomads (1000 BC), for instance, who already practised stockbreeding alongside cultivation, changed over to sedentary rice cultivation with cattle and horse breeding as early as 500 BC, in many areas. This stage has been maintained in many parts of southern Asia up to the present day; only a few groups (such as the Todas of the Nilgiri, for example) lived primarily from their buffalo herds.

Thus, two things led to pastoral nomadism: either partial nomads were forced for various reasons to live in dry regions where cultivation would be an inadequate basis for existence; or the population enjoyed a nomadic life which involved trade and marauding expeditions, so that they separated themselves from farming communities. These developments could take place at very different times and are by no means linked to definite phases of evolution (Dittmer, 1965, p. 20).

Even more important than the special case of nomadic pastoralism is the further evolution of the stages of the peasant economy, which can only be hinted at here. The most important feature of a group which was more closely bound to the soil was the development of seigneurial forms of organisation, with a strong social differentiation of the agrarian population; the fact that nomadic herders often imposed themselves upon farmers was certainly not responsible for this. In the course of time various social structures developed, among which the decentralised feudal and the highly centralised monarchic-bureaucratic forms are of special economic geographical importance (Bobek, 1959, pp. 274–9; cf. also, with reference to this chapter, the contributions of the Viennese social historian O. Brunner). The latter and more rigid structure developed in irrigation cultures whose weirs and other installations really called for a division of labour. Irrigation schemes, however, only gained importance with the development of urban civilisation. The development of the oldest urban settlements[1] (e.g. in Egypt or the Inca Empire) took place during this phase. Within more feudal regimes castles, smaller manors and estates grew up as new foci of the cultural landscape, and had a very marked influence on the clan communities of the rural areas.

With the further expansion of urbanisation it was not a great step from the manorial system to the stage of rent capitalism. In many parts of the east this new type of economy emerged logically from the union of manor and town. The term 'rents' means the regular share which townspeople held, according to their various titles or claims, in the produce of farmers or tradesmen (Bobek, 1959, p. 280). In the rent capitalist philosophy, the peasant economy was divided up into a number of factors of production: soil, water, seed, equipment, draught animals, and human labour. If the urban dweller succeeded in acquiring control of these production factors by systematically plunging the farmer into debt, rent capitalism in its purest form was the result; that is to say, the farmer's share was limited to

---

[1] On the basis of recent excavations at Jericho we can date the oldest urban civilisations as early as the turn of the seventh millennium BC (Meckelein, 1966). In the New World temple centres were often the nuclei of urban development.

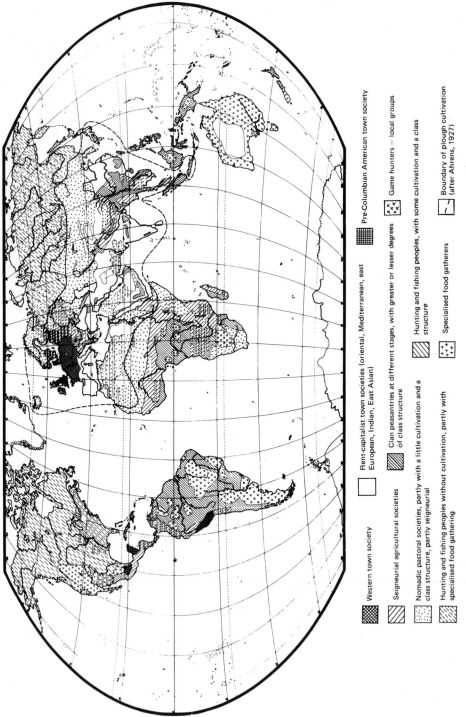

Fig. 3.1. The socio-economic stages of development in the world at the end of the fifteenth century (after Bobek, 1959)

the labour he provided. A typical feature of the oriental economic mentality was that the rent capitalist did not take part in production but was satisfied with pocketing his share of the output. Such rent capitalist systems were found also in the Mediterranean region, eastern Europe, eastern Asia and in the Spanish and Portuguese colonies.

In the Near East especially, the commercialisation of agricultural production has had serious consequences up to the present day. Farm output does indeed remain in the hands of peasants and tenant families, but commercial exploitation falls more and more into the hands of people belonging to an upper class, who take very little interest in the work of production. It is even possible, in the rent capitalist system, to get a higher return from the soil via tenants, many of whom exist on a bare minimum, than by employing free wage earners or slaves. This system extended into the tropics as well during the agrotechnical (irrigation, plough cultivation) and social developments of the pre-industrial period. A rent capitalist system such as this does not lack capital so much as the initiative of entrepreneurs willing to invest their money productively for the sake of the general economy. As a result, the state has usually had to introduce industry. A mixture of state and private enterprise is thus typical of the economic structure of such countries. The economic concept of rent capitalism has held back general economic development, especially in the Orient, up to the present day. Particularly in agriculture, private initiative hardly enters into the mentality of the moneyed classes, so that state measures have to replace private ones (Bobek, 1962a; Hahn, 1963).

In contrast with rent capitalism, which has existed for more than 4 000 years, the new era, which only began with the Industrial Revolution, has been called the stage of 'productive capitalism' (Bobek, 1959, p. 287). Whereas all the other changes we have discussed were evolutions rather than revolutions, the 'Industrial Revolution' of the eighteenth and nineteenth centuries, which started in Great Britain, certainly deserves its name. This new era in the history of mankind, characterised by an ever-increasing concentration of population, more intense division of labour, technical advance, the mechanisation of production and the application of scientific knowledge, also had a great effect on the development of tropical agriculture. The plantation economy, which was based on the exploitation of slaves belongs to the first stages of this development.

With respect to agricultural progress, this form of agrarian feudal capitalism, often characterised by strong central supervision, was a driving force in development. Examples which can be quoted are the Japanese landlords of the nineteenth century, or presentday Taiwan; the latter has developed rapidly under a strict Nationalist Chinese military bureaucracy—and American aid. On the other hand, 'bureaucracy' turned out to be a complete failure in many other Asiatic and African countries. As experience in Europe, USA, Japan and Taiwan has shown, rapid economic progress can be expected if, above all, peasants find themselves involved in a system of pressures and incentives, in which cash sales serve as an incentive and a pressure situation is brought about by taxation such as ground rent (Ruthenberg, 1967, 1971). In such a system an appropriate incentive is certainly better than too strict directives, which can achieve the opposite of what is wanted if control is too severe.

As in Europe, where agrarian evolution stretched over centuries, passing through many stages from the medieval three-field system to the scientific-ally based agriculture of today, there are now signs of change in many parts of the tropics, mainly to be seen in reforms of agricultural policy.

Some phases of development, with their important social as well as technical innovations (for example, the importance of manuring since Justus von Liebig), which have taken place gradually in the temperate zone, can be expected to be speeded up considerably in the tropics (for instance, in development from the primary into the secondary and tertiary sectors).

As everywhere, it is particularly important in the tropics to exploit local resources and labour to the utmost (Hutchinson, 1966, p. 10). It can be disastrous, however, to decide which are the best methods for tropical agri-culture according to European criteria. One example is the differing im-portance of organic manure, which has partly been abolished in European agriculture whereas it is indispensable in improving farming in developing countries with low wages.

Signs of increasing change in the agrarian landscapes of the world coin-cided with the beginnings of mechanisation and industrialisation. The first steam plough was used in 1860, followed by the threshing traction engine and the tractor (1918). Since the beginning of the twentieth century electricity has become more important in the countryside. Besides artificial manures and the tractor, which has almost entirely replaced the horse in western Europe and which, because it can be used on poor land, has re-duced the difference between easy and difficult soils, industry has offered further help in the form of pesticides and weedkillers. Technical develop-ments in agriculture in some of the younger overseas countries such as the USA and Australia, have progressed much faster than in the old regions of Europe.

In the second half of the twentieth century this type of economy, in vari-ous forms, has spread stage by stage, first from the towns[1] and later over large areas, and to a particular degree in the tropics and subtropics. This cultural change has often been initiated by adopting technical-scientific ideas carried to the most remote areas by modern means of communication (including telephone, radio and television). A worldwide wave of innova-

[1] In 1800 only 1·7 per cent of the earth's population lived in large cities (over 100 000 inhabitants); in 1900 only 5·5 per cent; by 1950 the figure had risen to 13 per cent and by 1960 to 17 per cent. Of these 17 per cent over half lived in approximately 100 cities of 1 million inhabitants, of which there were only three at the beginning of the nineteenth century (Meckelein, 1966, p. 9).

tion has been started off by applying well-tried methods of rationalising agriculture. In spite of many obstacles, these processes aim, admittedly in a superficial way at first, at a 'world civilisation' via 'one world'.

Attempts such as the 'Green Revolution' (cf. p. 190) to solve the agricultural and nutritional crisis of the tropics need a concerted effort from all mankind. Campaigns for literacy and education need large-scale assistance from developed nations. The building up of industries and communications, and the conservation of natural resources, need huge investments, which can only be achieved by an improved distribution of wealth between poor and rich countries. The fight against the population explosion by birth control, though important is not sufficient, as the people that have to be fed are born already.

It becomes apparent that agriculture, education, technological development, industrialisation, population control and environment are closely linked with the future evolution of mankind, in which peace is the main condition of survival.

4
The general patterns of agriculture

The spatial differentiation of agriculture

One of the main tasks of geography is to examine structural and functional patterns. Often structural and functional characteristics do not differ very much spatially, but have an obvious difference in aspect: a structural unit is defined by the homogeneity of its phenomena or by a characteristic symbiosis of heterogeneous objects; a functional unit is defined by the spatial effect of certain forces within it. Neither of these spatial forms is tied to a specific size, and they can be of very different dimensions.

In the spatial analysis of the tropics, various types of structure can be distinguished. For example, in physical geography the terms 'ecotope' and 'vegetation formation' are used in a structural sense in physiognomical and plant-sociological classification, even if there is an ecological component. One can think, for instance, of the steep slopes of tropical inselbergs or the typical association of tropical riverine forests and savannas in a region characterised by its climate, relief and vegetation.

In the sphere of culture it is more difficult to distinguish basic units. Structural units in human geography can range from types of social community to the economic structural areas in which we find certain sizes of agricultural enterprise or similar property systems: examples are the plantations of Central America, the agricultural regions of Nigeria (see Fig. 4.1), the parts of East Africa where property is divided on inheritance, or countries with tenant systems of the kind described under oriental rent-capitalism. From the point of view of production, one can pick out spatial types with a strong tendency to one kind of output (monocultures) or economically heterogeneous regions—which, however, exhibit within their boundaries a uniform social structure (Otremba, 1958, p.372). Physically heterogeneous structural areas, on the other hand, may be uniform from the point of view of social or economic geography.

In contrast to structural areas, functional areas are intertwined by reciprocal effects and interchanging and complementary processes. Emissions from a centre between them can lead to the superimposition and overlapping of various zones of influence, which may be separated by neutral zones. A good example of a functional relationship is the urban

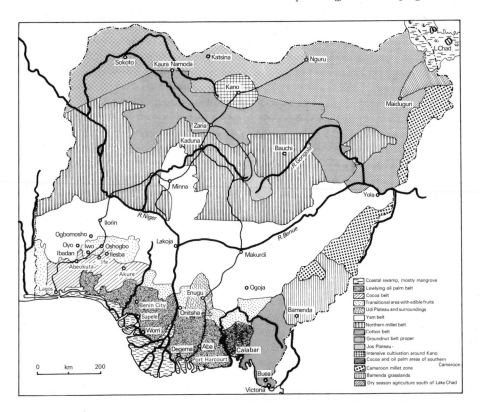

FIG. 4.1. An example of structural units in agricultural geography: the agricultural regions of Nigeria. In the arrangement of these regions one can see a sequence of climatic and vegetational zones, from the coast of Upper Guinea to the western Sudan (Nigeria including West Cameroons). (According to the *Sample Census of Agriculture*, 1952 and Manshard, 1963a.)

'sphere of influence' of a 'central place' on its surrounding region: the latter is supplied with numerous services and the settlement has a close connection with it (see Fig. 4.2). Such functional relationships can be observed in the most varied dimensions, from the isolated farm of the rain forest and savanna, with its gardens and fields, to the tropical metropolis with its zone of influence. At the national level these relationships are widened to constitute 'the market', and internationally to comprehend the polarity between core and periphery, that is to say, the tension between the industrial nuclei and the tropical lands which are their suppliers of raw materials and also their markets.

Besides occupying itself with questions relating to settlements and modern area planning, agricultural geography has concerned itself with functional spatial distributions by developing Thünen's basic ideas (1826). Waibel (1948) applied the thesis of 'Thünen's Rings' to the classification of the agricultural regions of Costa Rica; Pfeifer (1962) discussed similar ideas for Brazil. In so doing, old notions had to be adapted to the peculiarities of tropical agriculture and to modern transport developments, if they were to be valid for contemporary regional systematisation.

Other agricultural geographers (Van Valkenburg and Held, 1952; Grotewold, 1959; Chisholm, 1962) have also discussed the concentric rings of agricultural land use in Thünen's hypothetical state. This model has been extremely fruitful, although several theoretical revisions have been proposed (Garrison and Marble, 1957; Harvey, 1966). Future trends in this field have also been outlined by Henshall (1967). Other general models of agricultural activity have been formulated by pioneers of agricultural geography such as Baker (1926), Jones (1928–30) and Van Valkenburg (1931–36). Work on experimental and conceptional models has been well summarised by Henshall (1967); most of this work, however, has dealt with extratropical latitudes. In a model of tropical agriculture that part of the world lying between the Tropics is taken as universe and the basic types found within this zone are considered as sets. If as basic types of production, arable or pastoral, and methods of production, subsistence or commercial, are taken, four basic sets result. This arrangement of sets can be shown diagrammatically.

It is almost superfluous to state that both structural and functional patterns are only momentary representations which can change rapidly as a result of the dynamics of the forces at play. According to Wagemann's 'alternation law' (1943), population and type of economy often provide the impulse for development in a given economic region (Otremba, 1969). When an increase in population takes place, the agricultural area may either be expanded (e.g. by clearing land, establishing new settlements or emigration, in which case exploitation of the land remains as extensive as before) or new and more intensive economic methods may be introduced or developed on the spot, in order to be able to feed a larger number of people within the same area. Thus, an increase in population density does not necessarily imply decline. It is true that food supplies become more difficult and elbow-room more limited, but it is precisely such a development that can lead to dynamic expansion or even to industrial take-off in Rostow's sense (1960, 1971).

Whereas in most western industrial countries the present spatial structure has come into being without any overall central planning, so that spatial planning has been corrective rather than anything else, in many tropical countries central governments are largely responsible for planning development in general. One of the foremost fields of interest at the moment is an attempt to understand the relationship between town and country.

One of the most essential tasks, both theoretical and practical, of geography and spatial planning is to carry out research into questions of structure and function. Especially in the lesser-known and rapidly changing

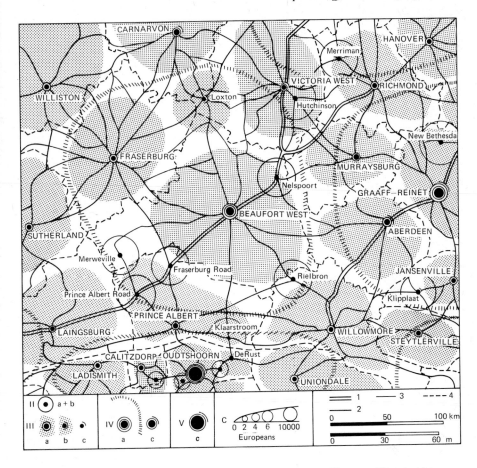

FIG. 4.2. An example of subdivision into functional areas: the Karroo of South Africa (after Carol, 1952)

II = Places with second-degree central services (complementary area shown schematically)
III = Places with third-degree central services (core zones shown)
IV = Places with fourth-degree central services (boundaries of core zones shown)
V = Places with fifth-degree central services (complementary areas not shown)
a = Full central place type
b = Semi-central place type
c = Subcentral place type
C = Central district, circle proportional to the number of European inhabitants
1 = National highways
2 = Main roads
3 = Secondary or link roads
4 = District boundaries

agricultural regions, it is necessary to check again and again the social and economic relationships between regions.

In this context, terms like 'active' and 'passive' areas (*Aktiv- und*

21

Passivräume) have been introduced into geographical literature (since the writings of Supan, 1922 and Schmitthenner, 1951). These terms are applicable to larger regions (continents and countries) as well as smaller spatial units. Between the two extremes there is a whole scale of transition. Important criteria in defining 'active' areas include an increase in population from migration and a specially high activity on the part of the non-agricultural population (which also brings about a greater change in the landscape). In contrast to these dynamic areas of concentrated economic power, which in English literature are often referred to as 'economic islands' (Green and Fair, 1962), there are 'passive' areas whose population is decreasing (frequently through seasonal or continuous migration) and which show fewer structural and physiognomic changes. The development impulses of the 'active' areas encroach deeply on the 'passive' or 'neutral-indifferent' areas and trigger off changes there, too, so that we have to consider both spatial units in close connection with each other. Perhaps it would be better to use the term 'weak' or 'slack' instead of 'passive', because the population of such areas does not necessarily behave passively at all; it may commute to distant towns or practise seasonal migration; in such cases there are often striking structural changes, and the decrease in population may lead to a 'healthy shrinkage process'.

Natural landscape belts (Landschaftsgürtel) as a basis for a broad zonal classification of the tropics

The whole zone of the tropics can be subdivided into vegetation formations, mainly defined climatically. This classification forms an essential premise for understanding tropical economic types, because in the broad vegetation belts the influences of climate, water availability, soil and relief offer different bases for agriculture.

Six main landscape belts closely related to the ecosystems can be distinguished (see Fig. 4.3) in the American, African and Australian–Asiatic tropics (Jaeger, 1945, Troll, 1951, and others).

In the core tropics evergreen rainforests and wet forests are characteristic. The lowlands near the equator are characterised by high relative humidity, quite heavy precipitation spread throughout the year, and high temperatures with little annual fluctuation. In the many-layered forests the tallest tree giants reach heights of over 50 m. The luxuriant growth of the tropical rainforest, however, should not hide the fact that the majority of its soils become impoverished rather rapidly, for it experiences a biological system with a closed cycle of nutrient matter. The soil constantly receives nutrients at the surface through decomposition of the vegetation cover; these nutrients, on the other hand, are rapidly washed away by strong rainfall so that—especially when human intervention is marked—the rain forest zone

Number of humid (or dry) months	10–12 (0–2)	9–10 (2–3)	7–9 (3–5)	3½–6 (6–8½)	2–3½ (8½–10)	1 (11)	0 (12)
Mean annual precipitation in mm	Mainly > 2000 mm	Mainly > 1500 mm	Mainly > 1000 mm	750 – 1000 mm	> 400 mm	Under 400 mm	
Schematic graph of annual rainfall	Axim 2103 mm	Tafo 1658 mm	Tamale 1081 mm	Kano 846 mm	400 mm	200 mm	
Examples:	Rubber, Tropical timbers	Oil palm, Cocoa, Coffee	Yams	Cotton, Millet, Ground-nuts	Ground-nuts		
Typical economically useful plants	Rubber, Tropical timbers	Oil palm, Cocoa, Coffee	Yams	Cotton, Millet, Ground-nuts	Ground-nuts		
Simplified transect sketch							
Plant-geography terms (Manshard, 1960)	Wet evergreen forest (rain forest)	Partly deciduous seasonally green wet forest (monsoon forest)	Wet savanna (with galleried and riparian forest)	Dry savanna	Thorn-bush savanna	Semi-desert	Desert

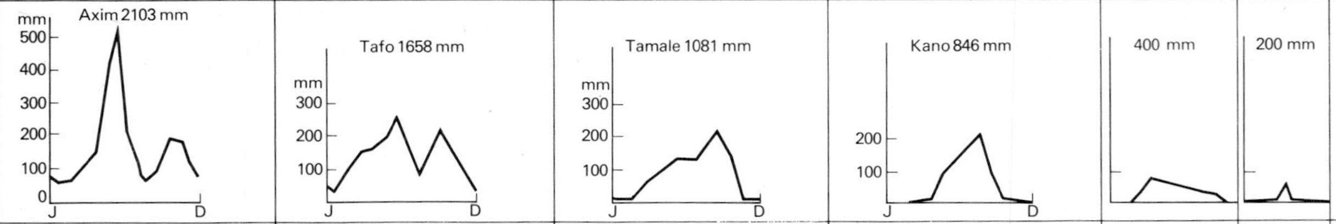

Fig. 4·3: Schematic summary of climatic vegetation formations (between the equator and the tropics of Cancer and Capricorn), taking West Africa as an example (Sketch: W. Manshard)

23

is only conditionally suitable for intensive permanent cultivation. Cattle and horse breeding face great climatic difficulties, and the danger of epidemics such as nagana.

Adjoining the rainforests both north and south of the equator we find seasonally green wet forests and humid savannas. There are no fixed boundaries between the individual vegetation zones; small stands of evergreens, riverine and gallery forests belonging to the humid zone are still to be found quite frequently here. However, there are already numerous species which shed their leaves during the pronounced dry period. Heavy rainfall causes leaching of nutrients and thus impoverishment of the upper soil horizons. Extensive laterisation is often an impediment to agriculture.

In the arid savanna zone (6 to 8·5 dry months) fluctuations in temperature and precipitation increase. Drought causes a pronounced rest period in vegetation growth and thus a hiatus in the agricultural calendar. Grassland often predominates. The open character of the landscape is often broken up by single trees, widely separated copses, or even xerophytic woodlands, momentarily green from some rainfall. Located between the climatic and agricultural aridity boundaries, on the frontier of precipitation-based cultivation, these dry savannas have frequently been greatly changed by man. Clearing by burning has destroyed part of the woodlands which previously existed.

Prolonged aridity (8·5 to 10 months) forces plants to adapt to climatic conditions. Reduced leaf surfaces, thorns, and ability to store water and assimilate it through the bark, are characteristics of vegetation in the thorn savanna (or bramble steppe) belts, where various forms of animal husbandry also predominate (nomadism in the Old World, stock ranches in the New).

Next to the arid boundary of the ecumene we find the zones of semidesert and desert (11 or 12 dry months). These regions are too dry to support a dense vegetation cover. Succulent plants prevail. Only during the short spell of rainfall in the semidesert is there some growth of grass and herbs. Periodic nomadism, hunting and collecting are the characteristic types of economy beyond the arid boundary of stock-farming.

The variety of tropical environments is reflected in the different agricultural zones, which tend to follow natural landscape belts; there is thus a close adaptation of types of agricultural economy and enterprise to basic vegetation patterns, themselves mainly defined by climate.

Agricultural geographers should pay particular attention to the three-dimensional climatic and vegetational zonation of tropical mountains to which Alexander von Humboldt referred in his travels in the tropics of the New World (1799–1804) and which Troll (1959) has represented in numerous comparative essays. Classic examples of such altitudinal zones in tropical mountain areas are the well-known belts in the Andes of South and Central America (*tierra caliente, tierra templada, tierra fría, tierra helada* and so on; see p. 149 and Fig. 4.4).

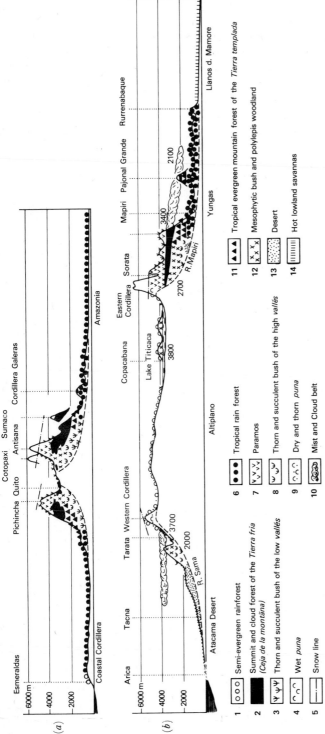

1 ○○○ Semi-evergreen rainforest
2 ■ Summit and cloud forest of the *Tierra fría* (*Ceja de la montaña*)
3 ᴪᴪ Thorn and succulent bush of the low *vallés*
4 ⌒⌒ Wet *puna*
5 --- Snow line
6 ●●● Tropical rain forest
7 ᴠᴠᴠ Paramos
8 ᴪᴪ Thorn and succulent bush of the high *vallés*
9 ⌒⌒⌒ Dry and thorn *puna*
10 ⬭⬭ Mist and Cloud belt
11 ▲▲▲ Tropical evergreen mountain forest of the *Tierra templada*
12 x x x Mesophytic bush and polylepis woodland
13 ⋰⋰ Desert
14 ‖‖‖ Hot lowland savannas

Fig. 4.4 Vegetation profiles across the Andes (after Troll, 1959, p. 44). The diagram shows two vegetation profiles: (*a*) across the equatorial Andes of Ecuador; (*b*) across the tropical Andes in the latitude of Lake Titicaca.

The equatorial Andes of Ecuador rise out of the constantly humid forests of the hot zone to both west and east, and thus show a more symmetrical vegetation profile. On both sides, at a higher level than the tropical rainforest of the *Tierra caliente* (6), there is the tropical evergreen mountain forest of the *Tierra templada* (11). In the *Valles* thornbush and succulent shrubs predominate (3 and 8). Summit and cloud forest (*Ceja de la montaña*) characterise the level called the *Tierra fría* (2), which higher up is bounded by Páramos.

In contrast to this symmetrical arrangement of vegetation in the equatorial Andes, southwards from Ecuador a sharp distinction develops between the dry western and wet eastern sides. The trade winds from the southeast bring orographic rain to the eastern slopes and give rise to a wet forest belt, whereas on the leeside western slope there is an intensified drought which gives rise to desert (13). The *puna* zone of the highland Altiplano lies midway between the two. From east to west the latter may be divided into the 'wet' or 'grass' *puna* (4), which still has a dense grass cover, and the 'dry' or 'thornbush' *puna*, where the grass cover is broken and thornbush and succulents appear (9).

We also find this zonation, although perhaps not quite so distinctly, in other mountainous regions of the tropics, for instance in the highlands of east and northeast Africa, in Malaysia and in New Guinea. In the high-lands of Ethiopia, for example, we can distinguish four altitude zones at short distances from each other: the *kolla* (up to 1 600/1 700 m), a hot belt with low precipitation; the more temperate *woina dega* ('wine country', at 1 600/1 700–2 400/2 500 m); the cooler *dega* (above 2 400/2 500 m); and the high region of the *tschoke*. In Ethiopia, the upper limit of settlement and crop cultivation lies at 3 800–3 900 m in Semia (north of Lake Tana) and at about 3 000–3 400 m in the south of the highlands (Kuls, 1958, 1963). The lower limit of settlement, in which crop cultivation can be carried on without irrigation, lies at 1 200 m in the southwest and 1 500 m in the north. Settlement and agriculture are particularly concentrated in the intermediate altitude zone of the *woina dega*, whose boundaries are marked higher up by frosts and low temperatures, and farther down by low precipitation. The actual and potential upper limits of cultivation and settlement lie rather closely together in overpopulated northern Ethiopia, whereas in the south there is more elbow-room. This contrast can be explained in part by the difference between the plough-using farmers of the north, who have been sedentary for a long time, and the communities in the south, who are stockfarmers or planters of ensete (which has an upper cultivation limit of about 3 000 m). The various cultivation belts also depend, naturally, on exposure and local climate, as well as on altitude and type of economy.

In understanding tropical agriculture, besides climatic and vegeta-

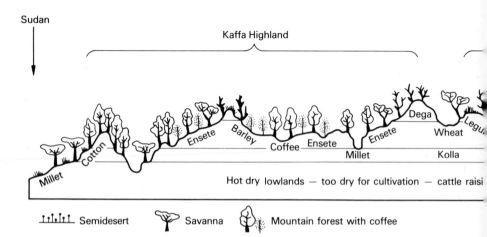

FIG. 4.5. Transect across Ethiopia (SW to NE). The most important vegetation levels and their agricultural uses become obvious in this cross-section (after Kuls, 1958)

tional conditions it is also very important to consider soil, in the broadest sense of the word, as the basis for agricultural production, as the supporter of nutrients and as the place where plants grow. Soil formation in the tropics, as in other climates, is affected by five main factors: climate, parent rock, biotic conditions, relief and time (see Jenny, 1941, Mohr and van Baren, 1954). Worthy of particular mention among climatic factors are high temperatures (up to 70° in the case of dark soils with strong absorptive capacity), which by promoting chemical decomposition accelerate the processes of soil formation. In the same way, high rainfall in the humid tropics and high evapotranspiration in dry regions are important for soil formation.

The effects of the space and time factors provide a great deal of information to both the pedologist and the geomorphologist, even though these effects are especially difficult to grasp in the tropics. In contrast to temperate soils, which frequently only go back to the Pleistocene, many tropical soils, with their vast weathering horizons, are very much older.

The term 'laterite' (Latin, *later* = brick) is frequently used to describe the soil of the humid tropics (the term being coined by Buchanan as early as 1807); but as Kellog (1950) points out, 'our conception of tropical soils has probably been influenced too much by laterite'. Laterites are characterised mainly by iron and aluminium oxides. They usually occur where relief is subdued and they are usually formed *in situ* on the underlying rock. Large areas of laterites are even as old as the Tertiary (Pliocene, Miocene; Finck, 1963, p. 48).

Many tropical soils, however, have no relation at all to genuine 'later-

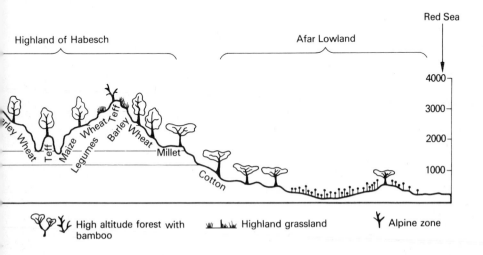

ite'. It is better, therefore, to speak of 'latosols', 'lateritic soils' or 'oxysols' when referring to the entire group of so-called zonal soils typical of the humid tropics. Other azonal types accompany these zonal soils—in particular, soils developed on alluvial river and marine sediments (in Southeast Asia, for instance) or on new material not laid down by water (regosols, on volcanic material, for example).

Compared with the zonal soils of the humid tropics, the zonal soil types of the dry belt are usually poorer in clay. Since more recent wind and high-tide deposits such as dunes and alluvial fans are more widespread, younger azonal (and also intrazonal) soils appear more frequently in the world's dry belts.

In comparison with the steppe soils of warm temperate latitudes, the soils of the dry savannas are usually poorer in humus and have a worse structure. When classifying desert soils it is best to follow the geographical divisions of types of desert (e.g. rock-, gravel- and sand-deserts, with their corresponding coarse or raw soils; Meckelein, 1959).

The so-called 'intrazonal' soils are of great importance for agriculture in the tropics; they are related to certain geographical factors and in a general classification can be placed between zonal and azonal soil types. They can be subdivided into halomorphous or salt-affected soils (such as *solontschak* and *solonetz*), hydromorphous or water-affected soils (for instance *gley*), and calcimorphous or calcareous soils. The widespread dark clayey soils (grumosols) or tropical black earths—such as the *regur* of India, the 'black cotton soils' of East Africa, or the *terres noires* of Zaire and Togo—can be included as well. These constitute one of the biggest soil reserves for future agricultural development in the tropics.

Milne's concept of the *catena* (1935), which was developed in East Africa, is regarded as an important contribution by pedologists working in the tropics. It can also be applied to agricultural and human geography. It is a principle of classification which works for both mapping and the classification of soil complexes. A *catena* (Latin = chain) is a sequence of soils, often completely alien from the point of view of a systematic classification but linked to each other by their topographic relationship, recurring regularly in comparable topographic locations, e.g. upper, middle and lower slopes (Fink, 1963, pp. 30–1).

In conclusion, we must stress the importance to mankind of the tropical areas, which have a high biological productivity, along with the greatest number and diversity of animal and plant species. The trouble is that we still do not yet know enough about them, how they work, or how to tap their resources without destroying them. Conditions in these areas differ widely from those in temperate zones so that we cannot just transfer our conventional agricultural methods there without modification. These observations will serve as starting points for an agricultural geographical classification of the tropics, which forms the substance of the following sections.

Bases for a broad classification of the tropics from the point of view of agricultural geography

The criteria on which a broad spatial classification of culture and economy can be based are varied. Several aspects could serve as the bases for a subdivision: for example, cultivation techniques (ploughing, hoeing, *milpa*, irrigation cultures), cultivated plants, or stages of economic evolution and social development. The aim of such a classification should be to delimit parts of the earth's surface in such a way that they are relatively homogeneous and yet distinctly different from other areas. Statistics are used for interpreting the economic conditions in such regions of the earth, but these have frequently been worked out for administrative units and only rarely do they apply to large natural regions.

The following are a few attempts at classification which seem important to our study of the tropics; though some are now merely of historical interest.

Classification according to agricultural zones

Engelbrecht's article (1930) on the *Landbauzonen der Erde* (Agricultural regions of the world) was a first attempt at an agricultural geographical classification of the earth's surface. It was based on the distribution of the various dominant cultivated plants (different types of grain, commercial crops providing raw materials for industry). It is interesting, in this latter context, that he excluded plantation crops because 'it is not infrequent that they are subject to abrupt and repeated changes' (p. 288). In describing the 'cultivated zones between the tropics of Cancer and Capricorn', which are of prime interest to us, lack of sufficient and accurate statistical evidence caused Engelbrecht to use as examples India and Java—densely populated and highly cultivated tropical countries. According to the dominant cultivated plants, he subdivided the tropics into two general agricultural zones:

(*a*) Tropical rice zone: in its distribution this coincides to a large extent with the natural vegetation zone of humid hot rainforest. Although Engelbrecht was aware of the various degrees of intensity of rice cultivation in the several culture areas of the tropics, he nevertheless classed them altogether in a single cultivated zone, because of the similarity of their environmental conditions.

(*b*) Other tropical agricultural zones: in their distribution these are mainly tied to the seasonally humid climate of the tropical savannas. They are characterised by two kinds of millet. In the wet savanna areas sorghum predominates, while with increasing drought it is replaced by spadiceous millet in particular.

One can see from the above that in Engelbrecht's rather simple classification agriculture was essentially conditioned by climate.

Classification according to suitability for cultivation

In his *Bonitierung der Erde auf landwirtschaftlicher und bodenkundlicher Grundlage* (Assessment of the Earth on agricultural and pedological bases) Hollstein (1937) distinguished six cultivation zones of varying potential:

(*a*) Cultivation of crops with different warmth requirements: in this region two harvests per year can be obtained from the same land. In summer, plants which require heat are cultivated (rice, millet, maize, beans) and in winter plants with low thermal requirements (wheat, barley, peas).

(*b*) Cultivation of 'warm' crops in summer: only one harvest of tropical grains possible during the summer rainy season.

(*c*) Cultivation of two field crops with equal warmth requirements: increasing rainfall allows two harvests, provided soil conditions are favourable. However, only crops with a short vegetative period can be used.

(*d*) Uninterrupted cultivation of 'warm' crops: in the constantly wet areas of the tropics, sowing and harvesting can succeed one another without interruption, with appropriate management (maize, rice, root crops, bananas).

(*e*) Cultivated areas in low-latitude highlands and mountains experiencing summer rainfall: highlands above 1 000 m have to be considered separately because they often have completely different yields from adjoining lowlands. With irrigation, double cropping is possible.

(*f*) Areas which cannot rely on rain for cultivation: climatically too dry and hot to be used for agriculture. Some of these areas are used for stock-rearing.

In representing the cultivated areas of the earth cartographically, Hollstein not only depicted their present use but tried to indicate their potential. Only if some of these regions are fully developed and made fully productive could they provide the yields shown.

If we follow Hollstein's ideas, this would mean that geography should not only carry out studies of physical geography and regional ecology but also, using the techniques of cultural and agricultural geography, explore the limits of possible land use. In such a case, the sociogeographical background of the land's capacity ought to be looked at especially, since it is very often social factors that impede development.

While Hollstein (1937) stressed the great importance of food production in the tropical world, Visher (1955) and Chang (1968) are more reluctant in their appraisal of the potentiality of tropical climates for agriculture. Factors such as solar radiation, high night temperature, lack of seasonality,

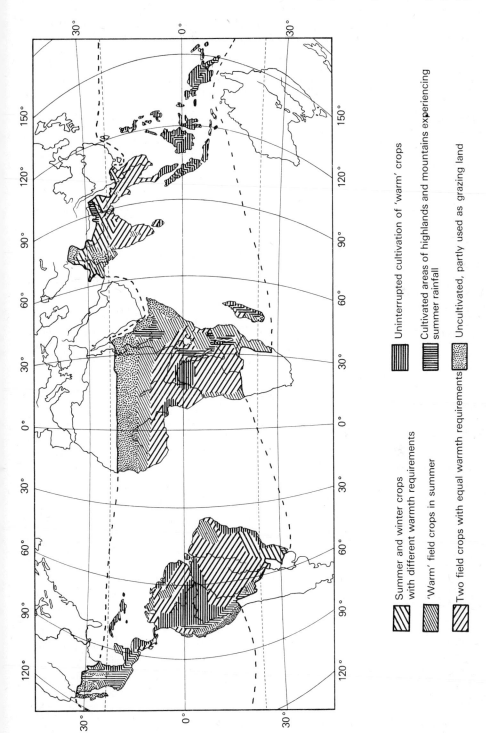

FIG. 4.6. Cultivated areas of the tropics (after W. Hollstein)

Summer and winter crops with different warmth requirements

'Warm' field crops in summer

Two field crops with equal warmth requirements

Uninterrupted cultivation of 'warm' crops

Cultivated areas of highlands and mountains experiencing summer rainfall

Uncultivated, partly used as grazing land

excessive rainfall and reduced potential photosynthesis are all factors limiting the possibilities of tropical lands (Gregor, 1972).

An attempt to classify the tropics according to dietary standards

Kariel (1966) drew up a spatial differentiation of the earth on the basis of food consumption, thus continuing the research which Hettner had already stimulated by several dissertations on *Speise und Trank* (Food and drink). From the geographical standpoint, in Germany Pfeifer in particular had earlier pointed out (1948) the importance of a scientific study of nutritional problems; Sorre had done the same in France (1952). In North America especially, groups of scientists from the University of Stanford (California) had pursued this line of research (e.g. Johnston, 1958; Jones, 1959).

Kariel's study is based on a qualitative classification which embraces both foods with a high calorific value (basic foods like grains, potatoes, sugar, fats, oils) and the main protein-containing foods (animal and vegetable). It describes average dietary conditions in different parts of the world and can be used as a starting point for a spatially differentiated classification; at the same time it suggests whether the various dietary conditions can be regarded as sufficient or insufficient. The study uses data showing the estimated average consumption per head per day of various types of food and does not, therefore, provide information about differences within a given country, which are related to such factors as climate and soils, economic status, and religious and national attitudes.

Kariel's general map is reproduced in Fig. 4.7, showing the dietary patterns of the world. The key is arranged in groups of dominant food grains: wheat, rice, maize, millet; within each group the major sources of calories are given, according to their importance in the diet. Although because of its global scale this map only offers a primary outline, nevertheless certain points emerge which help to characterise dietary conditions in the tropics. It is striking that tropical populations, despite all the variations in needs and tastes attributable to racial differences, and despite the variety of basic foods and the way in which they are prepared, nevertheless show one common feature in their diet: the predominance of carbohydrates and a deficiency in proteins, especially animal protein. The diet is mainly vegetarian, for even the demands for proteins and fats are generally covered by vegetable sources, which are, moreover, often inadequate, as, for instance, in the highlands of Central America, including Mexico. Animal protein in the form of meat is frequently merely an occasional addition. Fish is consumed only near coasts or rivers. Milk and dairy products are only consumed in any amounts worth mentioning by stock-farming populations. In the tropics, animal foods constitute as little as 4 to 5 per cent of the total calorific content of the diet. Almost everywhere, the population is capable of covering only the mini-

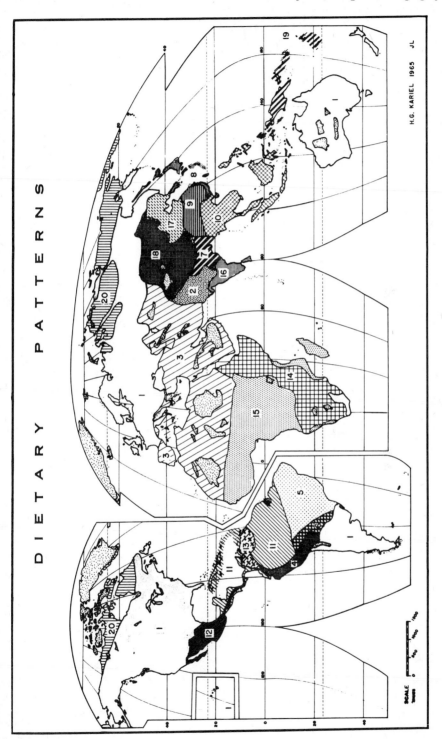

FIG. 4.7. Basic dietary patterns of the earth (after Kariel, 1966)

Major sources of calories:	Important sources of protein:
1. Wheat, potatoes, sugar, meat, fats, oils	Beef, pork, mutton, dairy products
2. Wheat, millet and sorghum, barley, rice	Dried beans, dried peas, chickpeas, lentils
3. Wheat, maize, barley, rice, fats and oils	Beef, pork, mutton, dried beans, chickpeas
4. Wheat, maize, barley, potatoes	Dried beans
5. Wheat, maize, rice, sugar	Beef, dried beans
6. Wheat, maize, cassava (manioc)	Beef, dried beans
7. Rice	Dried beans, dried peas
8. Rice, wheat	Fish, soyabeans
9. Rice, maize, sweet potatoes	Pork, fish, soyabeans, peanuts
10. Rice, maize, sweet potatoes, coconuts, cassava	Fish, soyabeans, peanuts, dried beans
11. Rice, maize, bananas, yams, cassava, sugar	Dried beans, dried peas
12. Maize	Dried beans
13. Maize, wheat, potatoes	Beef, dried beans
14. Maize, millet and sorghum	Dried beans, dried peas, chickpeas, lentils
15. Millet and sorghum, maize, rice, yams, cocoyams (taro), sweet potatoes, cassava, bananas	Dried beans, dried peas, peanuts
16. Millet and sorghum, rice, cassava, coconuts	Fish, dried beans, lentils, peanuts
17. Millet and sorghum, wheat, maize, potatoes	Pork, mutton, soyabeans, peanuts
18. Barley	Dairy products, mutton, goat-meat
19. Cassava, yams, cocoyams (taro), bananas, coconuts	Fish, pork
20. Animal fats, wheat	Fish, local game

Because of the scale to which the map is drawn, only a rough outline is possible, lacking many details. This is specially noticeable in those parts of the tropics which have been less intensively studied.

The following sections of the key are important for the tropics (including parts of the fringes of the tropics): 2, 3, 4, 5, 6, 7, 10, 11, 12, 13, 14, 15, 16, 19; they can be further illustrated by a few words:

2. Northern India and West Pakistan: Wheat is the major food grain. In addition there are millet, sorghum, rice and pulses. Fishing is important on the Indus and along other rivers.

3. Wheat is again the major food grain in this area, which stretches from southern Europe and the Middle East to Central Asia and into the African fringes of the tropics in the Sahara and western Sudan. Fats and oils are important sources of calories. Maize, barley and rice are additional foods. Dates (Libya, Tunisia) and figs (Algeria, Morocco, Tunisia) are among the staple products of North Africa. Meat is less important, its consumption decreasing from north to south. Religion prohibits the consumption of pork in the Islamic countries. Pulses are the major sources of protein. Fish is important in the diet near coasts and along rivers.

The diet of nomadic groups in the Sahara, North Africa and Arabia differs strikingly because of its one-sided dependence on animal herds, so that foods other than meat and milk become less important.

4. The dietary basis in the Andean highlands (Ecuador, Peru, Bolivia) resembles that of section 3. Important food grains are wheat and maize, and to a smaller extent barley. Potatoes and pulses (beans) do not fall far short of wheat in importance, but the consumption of fats and oils decreases.

5. Brazil (outside of the Amazon basin): wheat, maize and rice are almost equal in importance as basic foods. Besides beans, beef is important as a source of protein.

6. Paraguay and the lowlands of Bolivia: in comparison with section 5 manioc predominates. In addition, wheat and maize are major food grains, with dried beans and beef as main protein sources. Sugar consumption decreases.

7. North-eastern India, East Pakistan: in this and the following sections rice is the major food grain. Pulses are important as sources of vegetable protein. Fish is only of local importance.

10. Southeast Asia, southern China, Indonesia, Malaysia, Philippines: besides rice, maize and sweet potatoes, coconuts and cassava (manioc) are the main sources of calories (cassava especially in Indonesia). The major sources of protein are fish, soyabeans and peanuts. Meat consumption rises in Singapore and other towns (pork consumption by the Chinese population). Fish is only locally important (especially in Thailand).

11. The lowlands of South and Central America (e.g. in Ecuador and Peru), the Amazon river basin, Caribbean islands: rice and maize are the main food grains (millet in Haiti; cassava in Haiti, Cuba and the Dominican Republic). Sugar consumption is high throughout the area. Prior to shortages of foods due to the Castro régime, Cuba held the lead in the consumption of meat, fats and oils.

12. In the uplands of Central America, including Mexico, the diet is based rather disproportionately on maize and beans. There is little variety in the food. Occasionally it is supplemented by fruits, vegetables, dairy products or meat (beef and veal only).

13. The highlands of Colombia and Venezuela: wheat and potatoes are staple foods, besides maize. Generally the diet is more varied and meat consumption higher.

14. The highlands of eastern and southern Africa: the total consumption of cereals is rather constant but varies according to local conditions as to the proportions of millet (millet and sorghum especially in the Sudan and Tanzania), maize (Zambia, Malawi, Rhodesia, Kenya), teff (Ethiopia) and rice. Pulses are common. Meat consumption is relatively low. In the towns, especially in southern Africa, section 1 predominates, particularly among Europeans but also among urbanised Africans.

15. Africa south of the Sahara (including the Congo) and coastal lowlands of East Africa: total grain consumption is generally slightly lower than in section 14, varying regionally between millet, sorghum, maize and rice. Yams, cocoyams (taro), sweet potatoes, manioc (cassava) and bananas are important sources of calories. Liberia, Guinea and Malagasy lead in rice consumption. Wheat consumption is lowest in the Congo and in the rainforests of equatorial and western Africa; instead, the cultivation of manioc (cassava) rises; the consumption of meat and milk almost disappears.

16. Southern India and Ceylon: besides rice, manioc and coconuts, which are the main foods, there is a high consumption of fish on the coasts and an increased consumption of sorghum and millet inland.

19. Tropical Pacific islands: here as in many other areas, the diet has been strongly modified by contact with other habits. In the villages manioc, yams, taro and bananas, as well as coconuts, are still staple items of diet. Fish and pork are the main sources of protein. Breadfruit is locally important.

mum of its protein requirements. The consequences of a diet of such low nutritional value are often pointed out: unwillingness to work hard, proneness to illness, and so on. (cf. p. 135)

For the reasons mentioned earlier, many spatial differences in the dietary patterns of the tropics do not show up in Kariel's classification. This classi-fication was constructed by estimating food consumption from a com-parison of production, imports and exports (minus waste). It is not yet possible, from existing data, to make a worldwide classification on a proper statistical basis and relating to physical and social geographical regions.

Much preliminary work needs to be done in order that quantitative and analytical studies of geographical regions may prepare a method of con-sidering the different dietary conditions of the tropics in a more discrimi-nating way; after all, the way in which the human diet is obtained, together with its composition, also depend on anthropogeographical processes.

Classification according to cultural landscapes (*Kulturlandschaften*) and economic relationships

Jaeger (1934, 1943a) produced a classification of the earth according to cultural landscapes. It is mainly based on agricultural economic factors but also considers conditions of industry, transport and population. These indicate not only the type of economy but also the degree to which the cultural landscape is being transformed. Jaeger further subdivides these main types of land use according to their common cultural traits (the Ibero-American, Indian, Negritic and Malayan–Melanesian–Polynesian cultural zones, among others) and thus achieves a major regional typology. Otremba, in his attempt to classify tropical agriculture from the point of view of economic geography (1950, with map), puts the regions relationships with the world market in the foreground. Countries de-pendent on the world market, i.e. nations with a onesided economic structure which rely on world trade to sell their raw materials, are mainly dealt with in this classification, which also gives a preliminary impression of the world economy types characterising the tropics. Not only countries' positions as regards economic policy and world trade are taken into account, but also their internal economic structure and geographical situation, so as to be able to judge all the better the 'spatial economic problems of the developing countries' (Otremba, 1962).

Other economic geographers (Boesch, 1947, 1964, Alexander, 1963, Grotewold, 1959) have also expressed their opinions about this problem, partly applying Thünen's ideas, partly referring to Engelbrecht's cultivated zones.

It is the present author's view, however, that many statements about the economic classification of the tropics are no longer sufficiently close to reality, and the older economic geographical concepts of the tropics are already disintegrating. This process of disintegration has altered more and more areas, especially in recent years. Political emancipation and industrialisation are progressing all the time and the tropical countries, like other former colonies, are gradually disengaging themselves from the interlocking relationship in which coloured people did all the manual labour and white people were the brains organising them.

These increasing changes in economic and political structure vary in degree from area to area, as a comparison of economic data shows. As compared with tropical America and monsoon Asia, Africa shows the highest quota of agricultural workers, the lowest quota of international trade, and a negative export balance of foodstuffs, livestock and beverages. Nowadays, instead of an increasing relationship, as regards economic structure, between countries of the same climatic zone, there is an obvious tendency for countries of the same cultural zone to become more and more alike in their economic structure. This fact only reflects the principle by which regions of similar physical structure become more and more individualised as they are progressively opened up.

Cultural and economic continents

Another broad kind of classification, from the point of view of spatial economy, would be a division into *Kulturerdteile* (cultural continents), *Wirtschaftserdteile* (economic continents; Kolb, 1963, 1966), *Kulturreiche* (Troll, 1966) and large *Wirtschaftskulturkreise* (areas with significant common cultural and economic traits)—terms which have been coined and used by various geographers.

A natural continent is usually topographically and morphologically defined, whereas a 'cultural continent' means a region of subcontinental extent 'whose unity is based on the individual origin of the culture, on the unique combination of natural and cultural elements which have formed the landscape, on the independent intellectual-social order and on the closely related historical process' (Kolb, 1963, p. 3).

In contrast to the six (or seven) natural continents, ten cultural continents can be distinguished whose area is not at all constant, either spatially or in time; they possess their own dynamics and they have influenced one another in the course of their historical development:

1. The Western cultural continent,
2. The Russian or East European cultural continent,
3. The Chinese or East Asiatic cultural continent,
4. The Indo-Chinese, Indo-Pacific or Southeast Asiatic cultural continent,

5. The Indian cultural continent,
6. The Oriental cultural continent,
7. The Negroid cultural continent,
8. The Anglo-American cultural continent,
9. The Indian-Ibero-American cultural continent,
10. The Austral-Pacific cultural continent.

To a large extent, certain economic continents can be associated with these cultural continents. Economic continents are large areas in which certain types of economy have developed. Politically subdivided to a varying degree, they may assume subcontinental size. Within them, endogenic and exogenic economic forces work together. Just like the cultural continents, all the economic continents which lie in the temperate or subtropical latitudes exert strong influences on the tropics and to a large extent affect the culture and economy of the countries situated there. This applies to the Western as well as to the Anglo-American economic continent, to the Comecon powers, and to China and Japan.

Apart from these global influences and reciprocal effects, the cultural continents which lie almost entirely in the tropical zone are of special interest.

The influence of Iberian civilisation has created a large area in the American tropics which cannot find a parallel in any other continent. Amerindian cultures have been largely destroyed by Europeans or they have altered decisively. Aboriginal populations have been repressed or have mixed with the intruding Europeans and Negro slaves imported from Africa. Early separation from Europe (at the beginning of the nineteenth century) was disadvantageous for the development of the economy. It is characteristic for many Latin American states today to show a dualism in their economy, between a small group of large landowners or industrialists and a mass of poor population without property. Also typical are spatial contrasts between intensively used areas where population is concentrated and large areas of land used only extensively or not at all.

Mexico and the Andean states are characterised by a contrast between densely populated highlands and thinly populated lowlands. Rich mineral resources gave rise to mining centres in these countries. There are hardly any large industrial regions, however, as the minerals were usually exported in a raw state. These countries are distinctly world-market orientated. Their agricultural production is sufficient for their own subsistence and they have developed cultivation for export. Many Central American states and the West Indies, too, are strongly world-market orientated, though at the cost of depending for some of their food supplies on the world economy. They mainly produce cash crops, destined predominantly for North American markets.

In contrast to the west, there are extensive lowlands in the eastern parts of the American tropics. A large part of central Brazil is still very little known and almost bare of human life. Economic activity is concentrated on a broad coastal strip. Luxury products (such as coffee) are grown on large plantations and sold on the world market. A good road network, denser population and ore deposits in the Brazilian highlands were important bases for the growing industrial areas in the south of the country. Thus Brazil is outstanding as one of the largest tropical countries and the most important unit of the Indian-Ibero-American economic continent.

The broad landscape belts of the African tropics between rainforest and desert—which show a distinct correlation with agricultural zones (see p. 23)—are split up by numerous national boundaries which go back to colonial times. Recent developments have led to even greater political disunity. Many of the territories which formerly were politically and economically more closely knit (for example French West Africa, French Equatorial Africa, British East and Central Africa, among others) have separated from each other.

With the exception of Liberia and Ethiopia, all the countries of black Africa were colonial dependencies up to very recent times, with world-market orientation of their economies imposed by European powers. Following an early phase of exploitation and the slave trade, economic returns were increased by the planned cultivation of new crops. Low population density over wide areas of the continent was mainly responsible for the meagre utilisation of natural resources and their relative unimportance on international markets. The development of certain *Aktivräume* or 'economic islands' stands out distinctly among large areas of economic stagnation.

In contrast to the African tropics, tropical Asia has for a long time been a densely populated cultural continent, whose development has not been so deeply influenced by Europe compared with the tropics of the New World. Characteristic of the Indian 'cultural and economic continent' is a constant race between food production and population increase. India in particular, because of its high population figures (547 million inhabitants, 1971), has become an important Asiatic market. Agriculture differs from that of tropical Africa because the plough and irrigation are used in rice cultivation. Export crops (tea, jute) are also cultivated on plantations. Rather large ore deposits have led to the development of mining and industry. Smaller industrial centres are dispersed all over the country.

Relatively thin settlement and favourable natural conditions characterise extensive areas of mainland Southeast Asia. Its transitional position between India and China has become noticeable in cultural influences, while economically powerful colonial middle classes (expatriate Chinese) have made an impact on the economy. Lowlands are intensively utilised by ricefields, whereas shifting cultivation using slash-and-burn methods prevails in the mountains. In Malaya the plantation economy plays a more important role. Important deposits of tin (for example in the Malay

39

peninsula) give rise to a mining industry. A low level of industrialisation and a subsistence economy are typical.

In the islands of the Indian 'economic continent', extremely densely populated areas contrast with others which are very thinly populated. Important types of land use are intensive horticulture in the overpopulated islands and shifting cultivation in the empty areas. Export crops enrich the agricultural scene. World-market involvement by these countries is stronger in the case of some products than the goods exchanged within the economic continent. This is generally the case with the majority of tropical countries (in Africa and Latin America too) which so far have served predominantly as economically complementary regions for industrial countries outside the tropics.

Summary: synoptic classification of tropical agricultural regions

The above discussion makes it clear that it is very difficult to express in map form an agricultural geographical classification of the earth. There are a number of agricultural geographical maps for individual tropical countries (see the national atlases) but a complete survey is made difficult by the rather large gaps in the tropics and subtropics where detailed information is lacking.[1] Above all there is a lack of maps of the distribution of types of economy and stages of economic growth, and types of landholding. Representation of the areal distribution of individual cultivated plants can be found among the thematic maps of many atlases.

The tabulation in Fig. 4.8, which is the product of discussions at the Department of Geography of the University of Giessen, can serve as a first synoptic classification of tropical agricultural regions. It shows the most important types of agricultural economy and correlates them with major vegetation types (and also with the number of humid and arid months).

For further synoptic classification the following agricultural areas could be singled out:

AGRICULTURAL REGIONS OF THE PREDOMINANTLY HUMID AND SEASONALLY HUMID 'INNER' TROPICS[2]
(*a*) Shifting cultivation and land rotation with bush fallowing (p. 52),
  (semipermanent cultivation (Ruthenberg, 1967) often with uncon-

---

[1] Boesch (1965, 1966) produced a comprehensive cartography of global production—an interesting agricultural geographical contribution to solving the problem.
[2] Naturally there are also dry areas in the 'inner' tropics (e.g. East Africa), in the same way that there are humid areas in the 'outer' tropical zone (e.g. southern China).

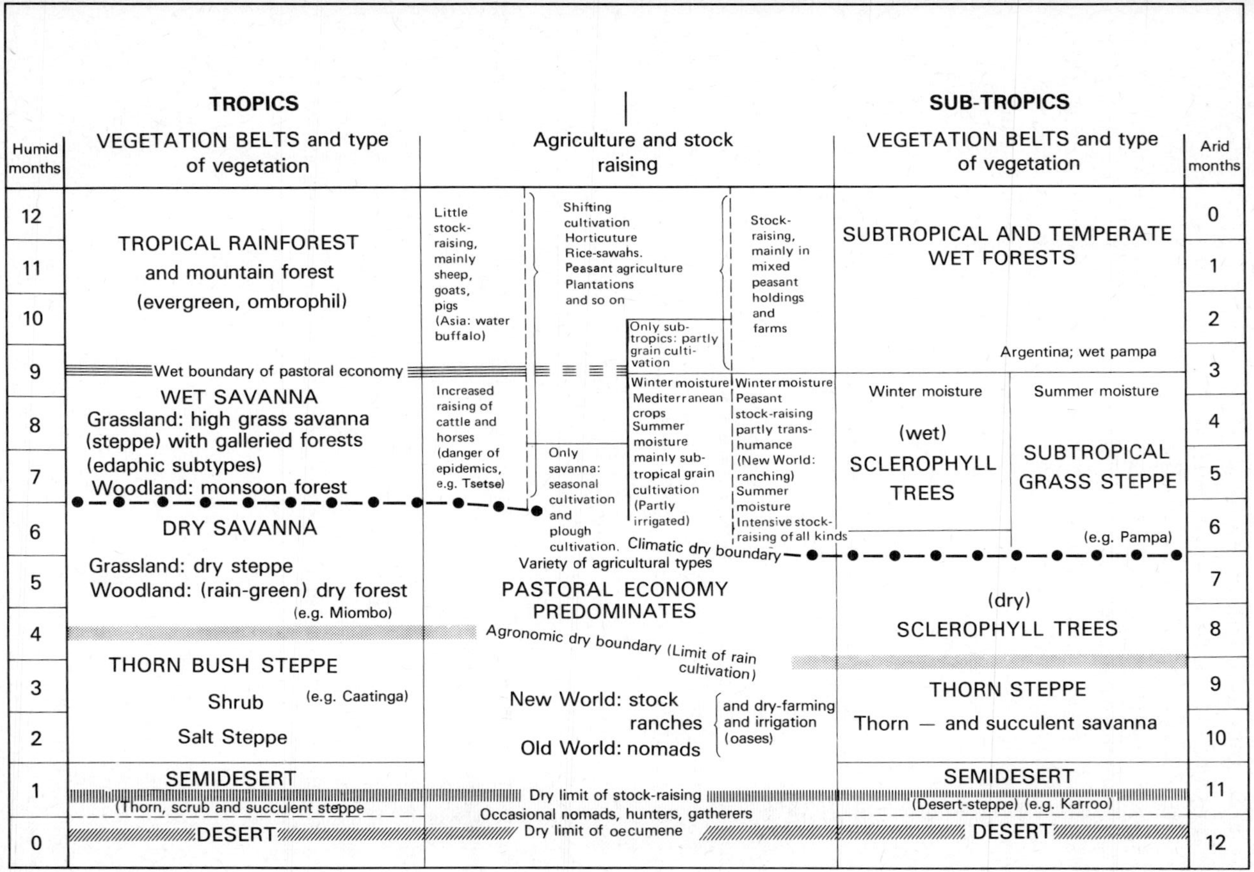

SUB-TROPICS

Arid months	VEGETATION BELTS and type of vegetation
0	
1	SUBTROPICAL AND TEMPERATE WET FORESTS
2	Argentina; wet pampa
3	Summer moisture
4	Winter moisture
5	SCLEROPHYLL TREES (wet) / SUBTROPICAL GRASS STEPPE
6	(e.g. Pampa)
7	(dry)
8	SCLEROPHYLL TREES
9	THORN STEPPE
10	Thorn — and succulent savanna
11	SEMIDESERT (Desert-steppe) (e.g. Karroo)
12	DESERT

Agriculture and stock raising

- Shifting cultivation / Horticulture / Rice-sawahs. / Peasant agriculture / Plantations / and so on
- Little stock-raising, mainly sheep, goats, pigs (Asia: water buffalo)
- Stock-raising, mainly in mixed peasant holdings and farms
- Only sub-tropics: partly grain cultivation
- Winter-moisture Mediterranean crops / Summer moisture mainly sub-tropical grain cultivation (Partly irrigated)
- Winter-moisture Peasant stock-raising partly trans-humance (New World) ranching / Summer moisture Intensive stock-raising of all kinds
- Increased raising of cattle and horses (danger of epidemics, e.g. Tsetse)
- Only savanna: seasonal cultivation and plough cultivation. Climatic dry boundary
- Variety of agricultural types
- PASTORAL ECONOMY PREDOMINATES
- Agronomic dry boundary (Limit of rain cultivation)
- New World: stock ranches / Old World: nomads — and dry-farming and irrigation (oases)
- Dry limit of stock-raising
- Occasional nomads, hunters, gatherers
- Dry limit of oecumene

TROPICS

Humid months	VEGETATION BELTS and type of vegetation
12	TROPICAL RAINFOREST and mountain forest (evergreen, ombrophil)
11	
10	
9	Wet boundary of pastoral economy
8	WET SAVANNA / Grassland: high grass savanna (steppe) with galleried forests (edaphic subtypes)
7	Woodland: monsoon forest
6	DRY SAVANNA
5	Grassland: dry steppe / Woodland: (rain-green) dry forest (e.g. Miombo)
4	THORN BUSH STEPPE
3	Shrub (e.g. Caatingal)
2	Salt Steppe
1	SEMIDESERT (Thorn, scrub and succulent steppe)
0	DESERT

FIG. 4.8. Types of agricultural economy in the tropics and subtropics (showing their relationship to the number of humid and arid months and to vegetation types). After Uhlig (1965) and Lauer, Jaeger, Troll, von Wissmann, Penck, Wang, Falkner, Jäzold and others.

41

nected agrarian landscapes; small population with only a few cultural islands)

    i. Permanently and seasonally humid forests (without lengthy interruptions in cultivation).

    ii. Savannas of various types as well as xerophytic forests with cultivation interrupted in the dry spells.

(*b*) Types of mainly permanent cultivation

    i. Horticulture (p. 83) (e.g. Java, Malaysia).

    ii. Dry-farming with plough cultivation and stock-raising (e.g. the Decan, highland cultivation in Ethiopia, the Bolivian Andes).

    iii. Irrigation practices with plough cultivation and stock-raising.

        (*a*) Terrace cultures (cf. p. 85) (especially rice, e.g. in Luzon, Java, southern China).

        (*b*) Wet rice cultivation in plains (e.g. Burma, Ganges valley, Kerala, Thailand, South and North Vietnam).

    iv. Plantation cultures; plantation and world-market orientated peasant economy (p. 99) (e.g. in Malaysia, Indonesia, Ghana).

AGRICULTURAL REGIONS OF THE DRY LANDS IN THE 'OUTER' TROPICS

(*a*) Irrigation agriculture on the fringes of the tropics (cf. p. 67) (e.g. Sudan, Senegal, Queensland).

(*b*) Livestock ranching (p. 111) (e.g. Angola, parts of southwest Africa and Australia).

(*c*) Nomadic pastoralism (e.g. Sahara, Somalia).

(*d*) Oasis economy with 'artificial' irrigation, along rivers and in isolated nuclei (p. 67) (e.g. Mali).

(*e*) Hunting and gathering economy (p. 183) (e.g. by Bushmen and Australian aborigines).

Bordering the tropics proper are the agricultural areas of the subtropics, which can be divided into those with winter moisture (mainly tree crops) and those with summer moisture (mainly rice, sugar cane and cotton). The position of the limits of cultivation (e.g. altitude, aridity, moisture) depends on natural, economic, social and technical conditions, so that one has to distinguish between actual, possible and profitable boundaries.

This broad classification of tropical agricultural regions is made more varied by certain special cases wherein both structural influences (mountainous areas) and functional ones (industrial regions, towns) are reflected. Many regions are mixtures of land used for agriculture and land used for mining, industry and urban settlement, whose function is expressed in the character of the environment. Structural contrasts between town and countryside are specially striking in developing countries, where broad transitional zones do not usually exist. But with increasing processing of agricultural products and the growth of secondary industries, closer relations between countryside and town are developing in many places.

Some socio-economic typologies of the tropical countries[1]

Since the Second World War attempts have been made, especially by economists, to define more precisely the term 'developing country', which has often been used too loosely. The common characteristics of these countries have been noted and the varied forms in which they appear discussed.[2] Since almost all the tropical and subtropical states come into the category of economically underdeveloped countries, one must mention some of the attempts at classifying them which may be revealing to the economic geographer. A more exhaustive examination would go far beyond the scope of this study, which only intends to give some hints in this direction, using particularly German literature.

The developing countries, in all their variety, can be classified typologically from quite different points of view. In the literature there are signs of classifications emerging which are based either on the developing countries possessing the classical means of production (capital, labour or population, and natural resources) or with the stress lying on cultural-historical or socio-historical characteristics. The latter group leads to a division of the developing countries into the really underdeveloped countries (some retarded regions of Africa), old civilisations with out-dated feudal structures (Near East, Far East) and so-called 'new nations' (South America).[3] In other attempts at a classification the actual economic

[1] In recent decades economists and sociologists have started to take an interest in the comparative typology of the economic structures of the tropical countries, i.e. in questions which are so important for agricultural geography. Since much of this work is little known among geographers, the following pages attempt a short summary, with appropriate references to the literature.

[2] The 'developing countries' or 'underdeveloped regions' were at first classified according to the size of their national incomes. All the countries in which the income per head was lower than in the USA, Australia or Western Europe (cf. UN, *Measures of Economic Development of Underdeveloped Countries*, New York, 1951, p. 3) were regarded as developing countries. Thus the measure of 'development' was often the rate of growth of real income per head in a given social structure.

Although income per head is not an ideal criterion of distinction between developed and underdeveloped countries, it is often used as such because it represents a quantitative unit—even if very often it cannot be determined very exactly—and because, moreover, it takes increase in population into account. In order to increase the validity of the income-per-head measure, bearing in mind increases in population, indices have been introduced which take into account consumption of calories or energy, distribution of income, level of education and standard of hygiene.

[3] Guth has tried such a classification of the developing countries from the cultural-sociological points of view (1957, p. 82 ff.). In a global survey he distinguishes: (1) Old civilisations with outdated traditional feudal structures (e.g. the countries of the Near East), in which there is lacking a middle class which would be willing to enforce development on a large scale by acting as leaders of an industrial revolution. (2) So-called 'new nations' (e.g. many countries in South America) which are characterised by a favourable relationship between the productive factors of soil and population and in which deposits of raw materials favour a more rapid economic growth.

systems of the developing countries are given prominence. In this case, distinction is made between countries with mixed socialist systems, market-based systems, and colonial regions whose economic systems are not of their own choosing.

Rostow (1960, 1971), whose model is one of the best known, classifies developing countries into different stages of economic growth by quantifying their varied characteristics. In this way, developing countries can be divided into four types, based on rates of net capital formation and net social product, expressed as percentages. In later studies Rostow postulates various likely chances of development by putting special stress on social structure. Included in the first stage of development are those political economies which are close to traditional social systems. Technical knowledge is small and the population hardly participates in economic and social life. The possibilities offered by modern science and technology are not at their disposal or are not used regularly and systematically, with the result that a large part of existing economic strength has to be devoted to agriculture. The prevailing social set-up, moreover, hardly allows much chance of promotion or upward social mobility. A kind of 'social fatalism' is shown in the fact that grandchildren cannot expect to achieve a position in life essentially different from that of their grandparents. Although the individual may be able to improve his lot, the social structure remains basically unaltered. A solution of the following problems could contribute to further development:

1. educating people to meet the needs of an economically and politically modernising state;
2. developing a modern administrative system for economic and political institutions;
3. reorganising agriculture with the aim of increasing its productivity;
4. constructing a transport system and opening up sources of energy;
5. applying modern techniques to the exploitation of natural resources, and thus achieving export surpluses;
6. transferring political power to leaders with a positive attitude to progress.

The second stage comprises societies and political economies in a transitional phase. A minimal economic and social infrastructure exists. The stimulus to economic and social takeoff comes either from within the society itself or from outside. One prerequisite for development is the realisation that economic progress is not only possible but that it is the basic condition for the realisation of objectives such as national prestige, personal gain and general well-being (examples: Iran, Iraq, Pakistan, Indonesia).

The third stage comprises political economies which are making economic headway (takeoff societies). A characteristic of these countries is that they have overcome old impediments and resistance to economic progress (examples: India, China, Philippines, Brazil, Venezuela). On the way to the fourth stage, the 'mature society' develops further its technical and entrepreneurial abilities, in order to produce a desired range of goods. A fifth stage of development is reached at the level of mass consumption, which is hardly ever found in the tropics but stretches from North America, Europe and Japan towards the tropical zone. This stage reaches its summit in high expenditure on national welfare and social services. According to Rostow, a population accustomed to prosperity and social security not only strives after money and goods, but a new scale of values can emerge which is beyond mass consumption (stage six).

In a criticism of Rostow's analysis of the stages of economic growth it has been said (Knall, 1962) that an automatic progression to higher stages does not necessarily happen with all developing countries, since these countries' opportunities for economic evolution vary considerably. It is naturally very doubtful, also, whether development actually takes place in the sense of 'organic' growth or whether under certain conditions some stages cannot be missed out.

In his *Social Strategy for Developing Countries* Behrendt (1965, pp. 59–61) gives a survey of types of developing countries. He distinguishes four groups according to the main systems of economic and technical organisation and according to standard of living:

1. Industrial Central Regions (as, for example, the northeastern and central parts of the USA, west and central Europe, including northern Italy). Highly industrialised, densely populated areas with a relatively high standard of living.
2. Young agricultural regions whose youthful (and even colonial) economic development is mainly based on agriculture, stock-raising and tertiary activities (examples: the Middle West and western parts of the USA, Canada, Australia, South Africa).

These first two types lie outside the tropical zone. Only the third category belongs to it:

3. Old, densely populated agricultural regions which have been economically developed for thousands of years and are mainly dependent on agriculture and stock-farming. These areas are characterised by high densities of population, and simple but labour-intensive production techniques. They comprise large parts of India, China, southern Asia, and also Egypt and the West Indies. The only exceptions in these regions are the plantation areas, which frequently show strong European influence.
4. Old, less populous agricultural regions, with a low density of population relying mainly on agriculture and animal husbandry, and with mostly simple and often rather labour-intensive production methods prevailing. In addition, large agricultural and industrial enterprises, organised either collectively or by private capital, are present to a varying extent. Large areas of the USSR, the Middle East, Africa (except Egypt) and Latin America are of this type.

45

According to this very rough classification by Behrendt, at present about 15 per cent of mankind would have to be regarded as (relatively) 'developed' and 85 per cent as 'underdeveloped' to varying degrees. It is beyond question that all these groups, no matter how we may strive to place them in space and time, contain a multitude of very different economic, social, cultural and political phenomena (Behrendt, 1965, p. 70).

Lorenz (1961) starts from a criterion of size and offers a classification into large, middle-sized and small countries, which he works out from the points of view of world politics, the world economy and economic policy. There are three main types.

1. Inherently important in world political affairs. Great potential of economy and domestic market. Little participation in world economic affairs. Extensive planned economic development projects.
2. Potentially important in world politics. Economy has some potential and domestic market offers opportunities for some sectors. Limited participation in world economy. Partial or moderate development planning.
3. Little potential importance in world politics. Insufficient domestic markets. Dependent on participation in world trade. Rudimentary or very few development projects.

Knall (1962) tries to go further than these classification schemes, which have as their starting point the present economic and political situation; he tries to estimate the developing countries' chances for economic takeoff, taking several factors into consideration: how far they are endowed with natural resources, size of population, religion, customs, traditions and mental attitudes towards development ('economic mentality'), and also their socio-economic and political institutions.

Knall's suggested division of societies into two groups, 'active' and 'passive', could also be applied geographically to active and passive regions (see p. 35). A passive population shows signs of active forces which might promote development, but cannot, however, overcome the passive factors which hamper progress. In this context we may mention as characteristics of the passive society:

1. a mentality with an inhibiting attitude to development; the majority of the population shows little interest in progress and is hardly, or not at all, prepared to exert itself in the cause of economic growth;
2. a mental attitude concerned only with covering day-to-day subsistence needs and with traditional modes of conduct;
3. great influence wielded by intangible factors which hamper development, such as religion, taboos and traditions;
4. as a consequence of these factors, institutional services which are underdeveloped or totally lacking; hardly any efficient agricultural or industrial development institutions; backward local administration;

an underdeveloped credit system; an absence of organisations capable of planning in the context of the overall economy; few opportunities for education and training.

With the help of these qualitative and quantitative determinants of mainly passive populations, various types of developing countries can be distinguished, and characterised as follows:

1. a poor agriculture, some potential resources, inadequate domestic capital;
2. non-existent agriculture, large natural resources, large amounts of domestic capital (Saudi Arabia, Kuwait);
3. an agricultural economy; few natural resources; little domestic capital. Apart from intensifying agriculture, the only industries which can be set up are those which are closely linked to consumption (Honduras, Cambodia, Niger);
4. an agricultural economy; natural resource potential, hardly exploited; little domestic capital; greater participation by the population in the exploitation of existing natural resources would be helpful (Vietnam, Egypt, Liberia, Guatemala);
5. an agricultural economy; existing natural resources being exploited; modest domestic capital (India, Peru, Morocco);
6. an agricultural economy; existing natural resources being exploited; large amounts of domestic capital (Venezuela, Iraq, Iran).

In countries with a mainly active population, people are more enterprising and more conscious about progress. Religious tenets, taboos and customs which hamper development play a minor part. Moreover, effective stimuli are provided by development organisations and projects, by an efficiently-functioning banking system and by other social and economic development institutions. Knall distinguishes the following types:

1. an agricultural economy; few natural resources; little domestic capital (Taiwan). Despite having few natural resources the population, which is keen on development, exploits the possibilities of agriculture and industry;
2. an agricultural economy; potential, partly exploited natural resources; modest domestic capital. The readiness of the people to work and progress, even though they have little personal capital, offers a certain guarantee to foreign investors, especially in the industrial sector, and thus a good chance for development (Israel).

Baade (1964) makes a rather onesided attempt to classify the developing countries from the point of view of their chances of economic takeoff, distinguishing five classes:

1. countries which still needed a lot of support a decade ago, but which have made full use of development aid.

47

2. countries with rich natural resources and stable governments, which were in a position to take advantage of their chances of development;
3. countries with notable natural resources and stable governments, which nevertheless could not cope with their domestic problems;
4. countries with a rather high population but scanty mineral deposits, which were able, however, to make profits from oil deposits and thus start developing themselves. Their chances are more favourable;
5. countries with a high increase in population but without mineral reserves or export products worth mentioning.

On the whole, Baade bases his work on the existing economic and political situation, stressing certain quantitative factors and paying little attention to the qualitative determinants of the development process (for example, the attitude of the population to economic progress; the influence of religion and tradition on the 'economic mentality').

Salin (1959) suggests yet another way of classifying countries according to their potential for development. He postulates an ultimate goal, which is usually modelled on the ideal 'American way of life', and asks whether economic development proceeds in a logical sequence (for example, from an agricultural economy to an agricultural and manufacturing one, and finally to an agricultural, manufacturing and trading economy), or whether it is possible to leave out certain stages. Salin believes in such a capacity to 'leap-frog', making it a function of components such as mineral resources, size of population, readiness to work and save, and structure of society, as well as the influence of institutions and religions. To these are added 'irrational components' such as the myths of self-determination, nationalism or communism.

Salin divides the developing countries into 'zones' of an agricultural condition, in which industrialisation is out of the question and agriculture is, and is likely to remain, the deciding factor, and zones with a potential for industrialisation.[1]

The zone with a permanent agricultural condition comprises countries whose chances of development lie essentially in agriculture. Within the zones with a potential for industrialisation Salin differentiates between zones with a passive population and not much domestic capital, zones with a passive population and unused capital, and zones with an active population and not much domestic capital. Whereas in zones with a passive population industrialisation can often only be achieved by an economic system employing force, countries with an active population have an advantage in that existing productive energies can be harnessed (for example, Israel).

Geographers, too, have expressed their views on the complexity of the social economics of underdeveloped countries. After thorough statistical analyses Berry (1961) has laid down certain basic patterns in several works;

[1] It is not intended to discuss the rather poor choice of the term 'zone' in this context.

he relies on economic, technical and demographic factors and compiles indices of them. Bobek (1962b) followed up Berry's work and, with the help of the material in the *Ginsburg Atlas* (1961) and the *Demographic Yearbook* (1960), he too has laid down indices or a ranking of individual countries as to their economic development. Thus, he picks out groups of countries which can be put together in regional units:

1. non-Soviet European industrial countries, together with countries overseas colonised by Europeans (mainly from Northwest Europe);
2. the countries of the Soviet block;
3. the southern European countries, together with some countries overseas colonised by people mainly of Mediterranean origin;
4. Latin American countries with strongly mixed populations;
5. the countries of the Near East and Islamic North Africa;
6. tropical Africa;
7. South and east Asia.

The Science Policy Division of Unesco (1973) has also attempted to construct a typology of countries according to their level of development. This theoretical planning model using social indicators was applied to a group of thirty-four countries totalling more than 80 per cent of both world GNP and population.

Although these statistical efforts should not be given too much value, they nevertheless enable us to make certain comparisons.

To sum up: it can be said that relevant information is forthcoming about the basic economic structure of tropical countries, from traditional economic analyses and by employing fairly modest techniques. Complicated programming procedures are needed, however, to compute future growth processes over longer periods. Only thus can we build with the necessary speed models which in spite of being schematic still possess sufficient flexibility. These models should provide the basis for a sensible development policy.

Discussion about the delimitation of 'socio-economic regions' is still active. Recently a lot of effort has been put into large-scale investigations in which representatives of very different specialised disciplines have worked together.

5
Tropical systems of cultivation

Agricultural systems denote the ways in which agricultural activity is carried out, and are usually divided into: primitive and less primitive gathering economies; cultivation by digging-stick, hoe and plough; and special types such as irrigation farming, horticulture and plantation agriculture. When one comes to employ these general terms in a strict sense and within geographical limits, as descriptions of types of production they become problematic. Hence Otremba (1960) suggested another kind of classification, according to aims of production and kinds of organisation. He distinguishes the following main economic types, which are socially defined:

(a) individual enterprise whose purpose is self-sufficiency;
(b) individual enterprise whose purpose is to supply the market;
(c) collective enterprise whose purpose is to cover national demands;
(d) cooperative enterprise orientated towards world markets.

Of course these types can be examined from different points of view. The kinds of land use (e.g. shifting cultivation, land rotation, permanent agriculture) can usually be associated with certain phases in the development of economy and society. According to the degree of market orientation (commercialisation), a distinction is made between subsistence economies and semicommercial types—in which self-sufficiency and supplying the market take place side by side on the one hand, with the fully commercialised enterprise on the other.

According to the constitution of the labour force and the use to which it is put, a division can also be made into family farms (with hired labour) and tenant farms, and these again can be classified according to size into full-time and part-time, small and large enterprises. If we use the land-holding system, we can point to private property (single person, small or large families), plantation companies and various kinds of cooperatives, and finally state collectives. Thus, a great number of criteria have to be considered for an agricultural geographical typology.[1]

Natural, economic, and socio-institutional conditions of agricultural production vary widely from place to place and over periods of time. In the process of adapting cropping patterns and farming practices to the conditions of location, more or less distinct types of farm organisation have developed. In fact, no farm is organised exactly like another. For the purpose of agricultural development, however, in order to devise meaningful measures in agricultural policy, it is necessary to classify farms according to their main characteristics (Ruthenberg 1971). In such a classification the following 'farming systems' can be distinguished for the tropics:

1. COLLECTING

The activity of the food gatherer is often supplemented by hunting and fishing. It is above all a subsistence economy and only seldom is gathering done for sale (e.g. gum-arabic in the Sudan).

2. CULTIVATION

This can be classified

(a) according to the kind of cultivation practised (e.g. rotations, crop-grass economy, systems with regulated ley farming);
(b) according to the intensity of rotation;
(c) according to the water supply (e.g. arable irrigation farming or dry-farming);
(d) according to cropping pattern and animal activities (e.g. different types of land use such as coffee–banana or rice–jute holdings);
(e) according to the tools used for working the land (e.g. hoe, digging-stick, plough);
(f) according to the degree of commercialisation: from a subsistence economy to mixed types and finally to strongly commercialised enterprises in which production no longer aims at subsistence. Often these are systems with perennial crops.

3. GRAZING SYSTEMS WITH GRASSLAND UTILISATION

These can be subdivided, according to the degree of residency, into full-scale nomadism, forms of transhumance or sedentary stock-raising and ranching. The various forms of crop-grass economy (ley farming) which is a rotation of cultivation and pasture, are very important in the tropics. The crop-grass economy is usually irregular: the fields turn into grass-

---

[1] The International Geographical Union founded a special commission to carry out this work. The task of this commission is to investigate various principles of classification, especially in their quantitative aspects. Cf. Kostrowicki, Helburn (1967) for a detailed further development of this agricultural typology.

land or bush and are then used as pastures. However, in some high altitude parts of the tropics, as for instance in Kenya, one may observe more regulated crop-grass economies with a systematic layout and use of the grass. Actual cultivation takes place on allotments specially set aside for it, year after year. Thus, grassland and pasture are already strictly separated from the arable fields. For didactic reasons the foregoing introduction is restricted to the presentation of some of the most important fundamental types, whose spatial distribution will be described with the help of examples.

Shifting cultivation and land rotation

Shifting cultivation and land rotation belong to the oldest and most original forms of land use. In the English terminology they are comprehensively called 'shifting cultivation' (Nye, Greenland, 1960). At the present day, more than 200 million people in tropical and subtropical regions still use these systems of cultivation. At an early date, scientists of different disciplines recognised and studied the extent and importance of economies employing shifting cultivation and land rotation. These monographs and essays, however, are usually restricted to certain regions or are simply attempts to investigate individual factors more thoroughly, such as soil conditions or management. Shifting cultivation and land rotation could fit harmoniously into the balance of nature for thousands of years, because of relatively low population densities. It was only the population explosion which made inadequate these land use systems, which have considerably modified the forest and savanna regions of the tropics.

Definitions and terms

As many terms occur in the literature, it is necessary to define them. Shifting cultivation is a type of economy in which both economic area and settlement are moved at certain intervals. Techniques of clearing the land, field patterns and settlement types can vary. In contrast to this settlements are not moved in land rotation; only the cultivated area is changed in a regular sequence, thus producing shorter or longer periods of fallow.

There are several forms transitional between shifting cultivation and the already semipermanent forms of land rotation. Therefore the term 'shifting cultivation', commonly used in the English literature, is often applied to both these forms and to all the intermediate stages (Conklin, 1954, 1957). Some authors, however, use the expression 'land rotation with bush fallowing' for *Landwechselwirtschaft*. Terms such as *Brandrodungsfeldbau* or 'slash-and-burn agriculture' really only refer to the techniques of clearing land for cultivation, which can be used in both shifting cultivation and land rotation.[1]

The literature contains many different terms for shifting cultivation which are valid only for certain regions. The following summary collects together some expressions (the form of spelling depends on the particular transcription used):

ASIA
ladang (Indonesia, Malaya)
jumah, humah (Java)
ray (Vietnam)
tam-rây, rai (Thailand)
hay (Laos)
hanunoo, caingin (Philippines)
chena (Sri Lanka (Ceylon))
karen (Japan, Korea, Taiwan)
taungya (Burma)
bewar, dhya, dullee, dippa, erka, jhum, kumri, penda, pothu, podu (India)

AMERICA (see also table 6.1)
milpa (Mexico, Central America)
ichali (Guadaloupe)
coamile (Mexico)
roça (Brazil)

AFRICA (cf. also footnote p. 57)
masole (Zaïre)
tavy (Madagascar)
chitimene, citimene (Rhodesia, Tanzania, Zaïre)
proka (Ghana)

In connection with this explanation of terms, it must be mentioned that these land use systems are not phenomena specifically belonging to the tropical and subtropical regions, as is sometimes assumed. It is very probable, indeed, that shifting cultivation, using fire to clear the land, is linked to certain stages of economic and social development, since it also

[1] The term *Landwechselwirtschaft*, which was coined by Waibel, was adopted by both Unesco (1958) and the IGU, in the form of 'land rotation'.

The term *Wanderfeldbau* (shifting cultivation) expresses distinctly the element of mobility in this type of economy. Other terms which turn up frequently in the German literature, such as *unsteter Brandrodungsbau* (shifting slash-and-burn agriculture) or *schweifender Wander-hackbau* (shifting migratory hoe tillage) are not always apposite, as slash-and-burn or hoe tillage are not used in every case. So far, whenever the term 'shifting cultivation' has been used, an explanation has had to be given as to whether this collective term was being used in the narrower or broader sense. Despite Pelzer's objections (1957), the present author would wish to retain the distinction between 'shifting cultivation' and 'land rotation', because it is important for the agricultural geographer. The often-quoted exceptions (e.g. that completely sedentary groups may practise 'shifting cultivation') only seem to confirm the rule. Worthy of attention is Terra's suggestion (1957, 1958) that we distinguish between shifting, semipermanent and permanent cultivation. The essence of land rotation is stressed perhaps more in the term *Landwechselwirtschaft* than in 'semi-permanent cultivation'. Frequent shifting of settlements has naturally declined a lot, especially in densely-populated areas.

The term *Umlagefeldbau* (*Umlagewirtschaft*) has also been used for 'shifting cultivation'. Ruthenberg (1967) reintroduced the term *Urwechselwirtschaft*, following the example of Aereboe, Brinkmann and Woermann. There are objections to this term, especially as the Prefix '*Ur*' could evoke wrong associations, cf. *Urwald* (virgin forest), *Urgeschichte* (pre-history), *Urlandschaft* (virgin landscape).

existed in Europe in prehistoric and historic times. Depending on the natural vegetation of the particular area, and on the length of the cultivation, fallow, or regenerative period, the rotation of crop and forest (e.g. in rainforests), crop and bush (e.g. in the wet savannas) or crop and pasture (e.g. in the dry savannas) can develop (Ruthenberg, 1965).

Often attempts are made to compare crop-and-forest rotation systems, still partly practised, in the temperate zone (e.g. the *Haubergwirtschaft* of the Siegerland, the Sartage-system of the Ardennes or the Anglo-Saxon 'Swidden' agriculture) with tropical forms of land use. This is only possible, however, if one makes due allowance, as the land tenure conditions are often quite different. In such comparisons, the north German systems of moor burning (*Moorbrandkulturen*) and even more so the north European system of forest burning (*Waldbrandwirtschaft*) are often quoted as examples of a primitive rotation economy (*Urwechselwirtschaft*) (Aereboe, 1920, p. 455 ff).

Before turning to a few regional examples, brief reference may be made to the typical techniques of cultivation which can be observed in all tropical regions. One of the oldest agricultural tools is the digging stick, which is still used today in many tropical areas alongside modern tools. It is used both by tribes which cultivate tubers (Papuans in New Guinea or New Britain) and by those who mainly grow grain (e.g. the maize-farmers of South America). Some people even speak of 'digging-stick cultures'; the ergological aspect should not be stressed too much, however, in characterising a given stage of economy. In tropical Africa and in some areas of Southeast Asia, the hoe has long been a favourite tool alongside the digging-stick (dibble). Indeed, the hoe in its various forms has been used as a universal tool right up to the present day. Moreover, hatchets, axes and chopping knives of all kinds (machete, cutlass, panga, parang) are used for clearing land.

Not only the tools but also the working methods resemble each other in the clearing of forests. When the trees have been killed by ring-barking and burning, and after they have been felled, the twigs and brushwood are piled up and dried, and then burned just before the rainy season begins. In the slash-and-burn system, root stocks are not usually cleared, and individual live or dead tree stumps remain on the fields. More valuable types of trees are often felled first. After the first rains, holes are dug with the digging-stick or hoe into the soft, ash-covered soil, and in these holes cuttings (particularly in mixed agriculture) or seeds are planted. Regular weeding is necessary until harvest time, as weeds shoot up rapidly. Moreover, damage by birds, rodents and game has to be combated. Insects and epidemic diseases, on the other hand, do not have such devastating effects because of the frequent changes of location of the fields, for example, where shifting agriculture alternates with plantations over extensive areas.

When comparing the rotation systems of the Old and New Worlds, despite all the similarities one comes across different land tenure systems again and again. In Latin America a great many people live independently on property which is not their own. The insecurity of these squatters (*intrusos, parasitos* and so on) has a hampering effect on development. In the tabulation in Fig. 5.1 an attempt has been made to arrange synoptically the types of economy which in Latin American literature are superficially described as *agricultura migratoria* and *agricultura transitoria*. In this summary there is an intentional emphasis on a genetic sequence (from left to right); some examples are given by way of illustration. The transitions are smooth, for shifting cultivation and permanent cultivation can alternate within a very small area. Colloquial expressions such as *roça* (land being cleared) or *milpa, conuco* or *capoeira*, indicate 'shifting cultivation' in the broadest sense of the word.

Regional examples

In Africa shifting cultivation and land rotation are even more important than in tropical America or Asia. For this continent, a whole series of studies on the theme exists. On the other hand, there are fewer works on America. For Southeast Asia the studies by Pelzer (1948, 1957), Dobby (1950) and Conklin (1954) are well known. Dobby estimated that in Southeast Asia about one-third of the total agricultural area was under shifting cultivation. Valverde (1971) gives a succinct account of shifting cultivation in Brazil.

Spencer (1966) has published a very thorough study of shifting cultivation in Southeast Asia. He presents these rotation systems in their manifold natural and cultural intricacies, stressing also the role of religion, ritual and magic—factors which many ethnological studies would do well to take into account—alongside those of agronomics, economics and techniques. In an appendix (F. pp. 204–12) he suggests a hierarchical classification for Southeast Asia, illustrating it with numerous examples. The emphasis does not lie only on historic-genetic aspects but also on types of change (uncontrolled, 'linear' and cyclical) and the permanency of settlement (seasonal, periodic or permanent). Proceeding in this way, it becomes clear that these rotation systems are closely connected with other types of cultivation and enterprise (e.g. horticulture), which will be dealt with in later chapters.

A great number of regional studies demonstrate that the changeover from shifting cultivation to land rotation has become a general rule. In the ultimate, but not the least important, resort this is due to population increase and the scarcity of land resulting from it. In areas where there is a transition to land rotation, accordingly as the period of activity has become ever shorter, soils have become much poorer than in regions retaining 'nomadic' forms of genuine shifting cultivation.

European colonists, too, partly adopted native agricultural methods. Waibel (1955) illustrated by examples from Brazil how small-farming settlers, in this case German, played their part in devastating the land through taking over slash-and-burn methods from the 'caboclos'. These

Shifting cultivation (semipermanent residency)		Land rotation (permanent residency)		Fully sedentary permanent agriculture
Free shifting cultivation	Restricted shifting cultivation	Simple land rotation	Improved land Rotation	With plough, manuring
Unregulated ('wild') rotation of field and forest	Unregulated field–forest or field–brushwood rotation (*rastrojo*)	Unregulated field–brush wood–bush rotation (*rastrojo, monte*)	Unregulated to regulated pasture–wasteland–(bush) systems, partly with cattle and horses, grass–fallow rotation with transition to permanent cultivation	cattle and horses

Regional terms and examples

Milpa (Guatales) system (e.g. Mexico, Costa Rica, El Salvador) Conuco-system (Venezuela)		Field–wasteland rotation economy (Costa Rica) 'Fixed' shifting cultivation Temporary lease systems (Colombia)	Bush–field–pasture systems (clearing-lease, Costa Rica) Bush–field-pasture systems (Colono system, N. Colombia) Field-pasture–wasteland system (Venezuela) 'Improved land rotation system' (S. Brazil) Regulated field-pasture economy (Brazil) Field-pasture rotation economy (e.g. Chile, Ecuador)
Roça system (Brazil) *Capoeira* system (Brazil)			

FIG. 5.1. Shifting cultivation in Central and South America

negative results certainly did not always follow, as the successful cultivation of tobacco by Dutch plantation companies in northeast Sumatra shows. On land which had already been deserted by shifting cultivators, and which was overgrown with *Alang-Alang* grass, the Dutch planted tobacco for a short time and then followed it with a nine-year fallow.

Despite the striking similarities of the systems of cultivation in different regions of the tropics, numerous individual features have to be noticed; these are due partly to cultural-sociological causes and partly to the natural environment. These differences are particularly observable in those regions where distinct ethnic groups live together in a small space. Malaya is a good example. The Chinese guerrillas who invaded the country in the 1950s planted ricefields in their own traditional way. British air reconnaissance could pick out these fields easily, since they were different from native ones, and this consideration made the Chinese give up planting rice themselves and force the Malayans to do it for them.

A whole list of other systems of shifting cultivation or land rotation is described in the ethnological and agronomic literature on tropical Africa, but for the most part these have local significance only. They are often similar in that land is cleared and cultivated, and then a shift takes place, but they differ quite a lot in detail. For further study of these rather little-known variants of the rotation economy, a few references are given.[1]

A number of special forms of shifting cultivation and land rotation are due to the unsatisfactory nature of soils. *Chitimene* (a clearing or a cleared field) is particularly interesting and is known in many variants in Africa. It was first described for Zambia (Richards, 1939; Allan, 1948; Peters, 1950; Trapnell, 1953).

In the practise of *chitimene*, bushes, branches and bark are collected from a large area and burnt on the fields. For instance, on the poor sandy soils of the Rhodesian high plateau, each family belonging to the Lala tribe clears annually about 7 hectares of forest timber; the wood is collected and burnt. The burning of this amount of wood is enough to provide nutrients to enrich 50 ares of land (5 000 square metres), on which eleusine (a kind of millet) is planted. According to calculations by Gourou (1956), this system of cultivation, with a fallow of twenty-two years, allows a population density of not more than 2·6 inhabitants per square km, provided that it is a purely agricultural subsistence economy.

According to observations by Allan (1948) in Zambia, the ash of 2–5 hectares of dry Miombo forest (mostly *Brachystegia* and *Isoberlinia*) is needed for the cultivation of 1 hectare of eleusine (millet). This process yields, with more or less security, 15 dz (1 500 kgs) per hectare, even on poor soils. The disadvantage lies in the great consumption of timber. Allowing for some

---

[1] *Aisa-Mambwe system* (Allan, 1965, p. 71); *Bantu system* (especially in Rhodesia, Malawi, Zambia, Zaïre) (Gourou, 1953, pp. 26, 31; Bartlett, 1956, p. 707; Tondeur, 1955, p. 67; Beguin, 1960); *Bemba system* (Zambia) (Richards, 1939, 1958, p. 106; White, 1963, p. 365); *Dinka system* (Sudan) (Burnett, 1948; Bartlett, 1956; Allan, 1965, p. 69); *Eastern Valley system* (Senga tribes in the Petanke District, Zambia) (Allan, 1965, p. 101); *Fipa system* (Allan, 1965, p. 139); *Katiri system* (southern Sudan); *Congo system* (Nicolai, 1961; Sautter, 1951, pp. 64–72; Wilmet, 1963a); *Luapua Valley system* (Zambia) (Allan, 1965, p. 142); *Ngoni-Chewa system* (Zambia) (Allan, 1965, p. 99); *Nuba Mountain system* (Sudan) (March, 1936, p. 79); *Nyamwezi system* (Tanzania) (Blohm, 1931, p. 117); *Tiv system* (Nigeria) (Briggs, 1941, p. 9; Bohannan, 1954; Meek, 1957, p. 149); *Zande system* (Sudan) (de Schlippe, 1956).

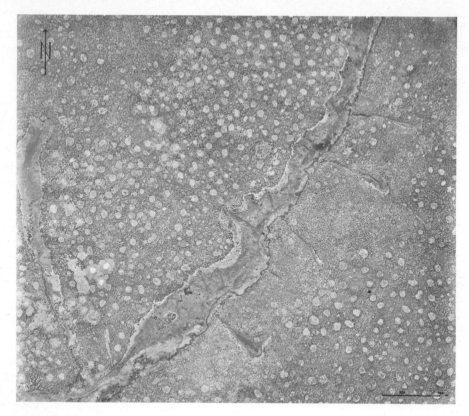

PLATE I. *Chitimene* gardens in different stages of Miombo regrowth in Chinsali District
(northern Zambia). Between the two seasonally flooded, grass-covered *dambos*
(Mwalese River) three new gardens can be observed (1965; scale 1:30 000)
(Permission: Surveyor General, P.O. Box RW 397, Lusaka, Zambia)

forest regeneration after about fifteen to twenty years, the nutritive content
of 30–100 hectares of forest is needed for the annual cultivation of 1 hectare
of eleusine (Ruthenberg, 1965). In other regions, the area stripped of
branches and brushwood is twelve to twenty times bigger than the fields
cultivated. The heating of the upper soil horizon on which the thick layer
of ash lies, is apparently beneficial for the soil structure and the control of
weeds.

Thus, the *chitimene* system is a rotation economy of field and forest, in
which the plant material won by clearing is transferred from a larger area
to a smaller one, in which it is used for the cultivation of agricultural
products.

Some authors (Hance, 1964; Allan, 1965) distinguish several variants of
the *chitimene* system such as, for example, the simple 'small circle *chitimene*';
its development, the 'Large circle *chitimene*', in which the clearing takes
place over a larger radius; and the 'semi-*chitimene*', in which an area in the

PLATE 2. *Chitimene* garden planted in third year with cassava. Kasama District, Northern Zambia

middle of the cleared 'island' is specially tilled and planted with millet. The tool used almost everywhere for cultivating not only the *chitimene* fields but also gardens, is the hoe.

The Bantu, Bemba and Dinka systems (see p. 57, f.n.) also resemble these types of field–forest rotation, wherein the total area needed is much bigger than the area actually cultivated.

As is well known, many African tribes even nowadays are unable to cultivate the heavy soils of the alluvial savannas (*dambos*) with the tools at their disposal. With the present rapid increase in population, the reclamation of these valuable alluvial soils by new techniques would offer a way out of the dilemma of increasing soil impoverishment and ever greater demand for food.

Further special forms of *chitimene*, as practised in south Tanzania and by the Dinka in the southern Sudan, have been described (Burnett, 1948). There the transformation of timber into fertiliser is achieved not by fire but by consumption by termites. In this Dinka system the fields are covered by a thick layer (50–60 cm) of bushes for about four months. As it is a purely biological process, the eating of the wood by termites, not so much nutritive material is lost as in burning. Within the cleared 'island' the tree stumps are simply left to the attentions of the termites, without any additional piling up or burning.

59

In Ghana there is the *Proka* system (Manshard, 1961a), a form of clearing land without using fire. Weeds and underbrush are removed and smaller trees felled, these then being left to rot on the spot or piled up in compost heaps in various parts of the field.

Finally, a further technique of rotating fields and grass is known in the southern Sudan; here the grass is not only burned off but killed. This method is also called *hariq* (Arabic: to burn off) and probably originated in Asia. In this method, after a close season of two to four years, the fire is laid with the help of old dry grass—not in the dry season but after the first rains, when the young shoots are coming up (Tothill, 1940; Burnett, 1948; Crowther, 1948).

With some Bantu groups (for example, in western Zaïre) an improved land rotation system brought about a reduction of the rotation area. This, on the other hand, led to the protection of fallow secondary forests before renewed burning (cf. the 'Bantu system', after Vandesyst, 1924; and Kuhnholtz-Lordat, 1939; Tondeur 1955), that is to say, a kind of forerunner of the forest reserves created in many parts of the tropics by the European colonial powers. A combination of land rotation with directed forestry and timber management suggests itself (see p. 110). Before burning, precious woods and timber trees are felled; and after the cleared 'islands' have been left, half-wild tree cultivation frequently remains. Many oil palms and cola trees in West Africa, the sugar palms of the Batak lands in

PLATE 3. Maize store used by peasants practising shifting cultivation in Zambia

Sumatra, or the orange trees and betelnut trees of Burma are examples of such an integration of land rotation and forestry, still usually rather loose; agricultural advisers of the former British colonial government recommended stricter management within the compass of a forest rotation (see p. 110), in the so-called *taungya* system.

Taungya is the term originally used for a 'shifting field' in Burma. The German forestry scientist Brandis (Stebbing, 1922, p. 376), who was in the service of the British, realised that the shifting cultivation system, which is so detrimental to timber resources, could possibly be rendered useful to the development of forestry. As early as the middle of the nineteenth century, teak and rice seeds were distributed to farmers. Two decades later, this rotation system had proved so good that teak trees could be grown very much cheaper than on commercial plantations. At the same time the *taungya* system was of educational value. The farmers no longer had to defend themselves in court for destroying the forest; indeed, they promoted reafforestation of the cleared land by sowing teak seeds. After clearing and before reafforestation the *taungya* fields are usually used for the intermediate cultivation of rice, tobacco and sesame, in the valleys, and of sugar cane, cotton and maize on the mountain slopes.

The Forestry Commission later introduced this system into British India, too. In the Siwaliks ('Bhabar'-alluvial cone zone, between the Ganges and the Jumna), for example, twelve new settlements were created with their help. The farmers were each given 0·3–1 hectare of woodland to clear and cultivate. During the first two years the whole area was cultivated, whereas during the next three years it took place between the sal (Shorea robusta) and teak seedlings which had been planted. The farmers promised to look after and weed the plantation in return for the seedlings supplied by the Forestry Commission. The term *taungya* was adopted for this practice, after the Burmese example.

In Thailand, too, the forestry authorities are deeply concerned to restrict shifting cultivation to the deciduous monsoon forest where there is no teak, in order to lessen the danger of a complete destruction of teak resources (Lötsch, 1958).

In contrast to the original native forms of shifting cultivation and land rotation which have been described at the beginning of this section, the following example illustrates how European influence sought to achieve an intensification and control of cultivation through strict organisation. A 'corridor (*couloir*) system' was practised in the Belgian Congo after trials which took place in 1935 and more intensively in the 1950s. Wilmet (1963a) and Tulippe (1964) give a good summary of the development of shifting cultivation into land rotation in Central Africa, and into improved forms of cultivation under European influence. The land was divided into parallel production strips 100 m in width. These strips, which were at different stages of cultivation or bush fallow at any given time, were then subdivided in such a way that each farmer and his family got their due acreage of property. This piece of land usually cut the long corridors at right angles and was composed of land of varying quality, at different stages of the cultivation cycle. This system, which could be modified according to local conditions, demanded strict central control, as it was carried out by the Belgian colonial administration (de Coene, 1956). A cleared corridor of this kind, on which the owner was obliged to practise uniform sowing, tilling, harvesting, etc., was shifted about within a wider area, which thus consisted of strips of land with the same crops or at the same stage of regeneration. In the course of time about 500 000 hectares were cultivated in this way, the so-called *paysannats indigènes*. In 1958 almost 200 000 African peasants were organised in such *paysannats* (Hance, 1964, p. 325). Such control of shifting cultivation (*Disziplinierung des Wanderfeldbaus*, Ruthenberg, 1965) had certain advantages. The proportion of cultivated and fallow land was under strict control. Periods of fifteen to twenty years without any cultivation were planned. In this way, soil exhaustion through too-short periods of fallow was avoided. One could specialise in the cultivation of a select few crops destined for sale. Also, the spreading of insecticides, weeding and improved crop sequences could be carried out more easily. Indeed, over a large part of the *paysannats* permanent cropping was instituted, without shifting cultivation entering in at all.

The Belgian example found imitators in neighbouring French Equatorial Africa where, in 1936, similar *paysannats* were created in the Central African Republic (formerly Ubangi-Chari); here, individual family properties were to replace the old traditional communal land rights. These measures, which were to introduce new cash crops at the same time, failed at first because of the passive resistance of the population, which was directed at the necessarily coercive measures adopted. Attempts were renewed in 1953, but in spite of initial successes the agricultural situation could not be essentially improved this time either. Many plantations began to revert to bush as the Central African Republic possessed neither the means nor the personnel to carry out the necessary agrarian planning (Hance, 1964, p. 295).

Some disadvantages of the corridor system are the difficulty of dividing the land in accordance with soil quality and the problem of estimating correctly the labour capacity of an individual family in distributing the land. The individual peasant had to follow a uniform sowing, tilling and harvesting programme. The supervision of these *paysannats* was very expensive, yet though a lot of land was needed, labour productivity was low. Thus, it is not surprising that once control by the Belgians was gone, this form of organisation collapsed almost everywhere, despite striking successes in production.

Resettlement in compact villages of populations which hitherto lived mainly in dispersed settlements (e.g. 'villagisation' and 'community development' in Tanzania) has intensified, for obvious political and social reasons, since many African and Asiatic countries achieved independence.

Israel is often quoted as an example here; the farm settlements of East Nigeria are another example.

Fields and settlements

In comparison to the dwellings of hunting peoples—primitive wind-and-weather shelters or simple semi-beehive or dome-shaped huts made of twigs —the semipermanent or permanent settlements of cultivating peoples are more clearly differentiated. In the original shifting cultivation system, settlements are mobile. It is during the land rotation stage that peasants gradually become sedentary. They lay out fields which are often several kilometres away from the settlements. In some regions (e.g. south Ghana) settlement locations have been fixed for several decades, or even for centuries. Occasionally, land rotation is even carried out from towns (e.g. among the Yoruba of west Nigeria).

When permanent nucleated settlements began to be formed, division of the land in both infields and outfields occurred simultaneously. This subdivision of the fields had a great effect on the further growth of settlement. When surplus population becomes too great and the journey to the fields too far, individual families split off and migrate from the centre to the periphery of the village lands.

On the periphery, in the zone of land rotation, small dispersed settlements are founded, usually 4 to 5 km from each other. Around each of these new settlement cores a cultivated area grows, modelled on that of the old village; different belts are formed, beginning with intensive garden cultivation and extending outwards to the extensive cultivation of the bush.

Frequently, huts which were used as shelters during harvest time formed the core of the new outlying settlement. The introduction of the bicycle made it possible to make use of land even further away from the village, and to build huts there. The additional development of certain functions in the original village, however (church, school, clinic, and so on), and the construction of more solid houses, can hamper the creation of new outlying settlements (Manshard, 1962). Formerly, locations near to crops and water and protected sites (mountain ridges, meanders, and so on) were preferred for settlement. At the present day there is a tendency all over the tropics to settle near roads and thus be more closely connected to markets and better job opportunities.

In connection with this trend one can observe urban building elements being intermixed with, and strongly influencing, rural house types. Thus, one can no longer sustain the regional differentiations which were formerly postulated about hut types (e.g. the beehive hut in the interlake areas of East Africa) and house types (e.g. the cone-shaped roofs of the savannas of the Cameroons and the gabled roofs of the rain forest).[1] A special case is

---

[1] A hut is usually a dwelling place in which the wall and the roof run into one another. In a house, on the other hand, the roof is distinct from the wall and forms a separate element in construction.

found in the lake villages (*Pfahlbauten*) of Southeast Asia, Oceania and Dahomey. Here, transformation and adaptation to modern conditions is more difficult to carry out, because of technical architectural reasons.

As already mentioned, certain basic types of cultivated area can be found in the agricultural systems of the temperate as well as the tropical zones, and in permanent cultivation as well as in land rotation. For instance, a more-or-less circular arrangement of the cultivable area of a village or individual holding is frequent. The intensity of cultivation gradually decreases in concentric circles, from the gardens alongside the houses to the infields and finally to the outfields, which are often widely scattered. The inner belts are usually characterised by intensive permanent cultivation, often with manuring by waste material and dung. The house gardens are sometimes no more than a few metres in width and they are protected by hedges or low stone walls from livestock and game. Occasionally these gardens form an enclosed area around farmsteads and settlements; often the inner belt is irrigated. Moving outwards, there is a second belt which is no longer so intensively cultivated; its fields are only occasionally manured by putting cattle on them after the crop has been harvested. In the third (outer) belt the small and scattered cleared 'islands' penetrate into the savanna or forest. They are only utilised extensively in a land rotation system. The greater part of employment is in the infields (permanent cultivation). An arrangement of fields such as the one described can be found, for instance, in Guinea-Bissau, in the western Sudan and in East Nigeria (Fig. 5.2 shows an example from the Gambia).

The rights of use and the distribution of property differ from group to group even within the same area. The Bouna of Upper Volta, for instance, divide their land into several sectors, with each sector containing different types and qualities of soil. Utilisation of the inner sector is reserved to the large patriarchal families, who work it separately. These old-established families have a monopoly on the manured land. Immigrants from elsewhere can only acquire land in the middle and outer zones (Izard, 1958).

The basic division of village land into infields and outfields was also common in some parts of Europe before artificial manure was introduced. At that time, after the soil of the infields had been exhausted, the outfields had to be put under more intensive cultivation. In Europe the clear division between infield and outfield has been proved by the study of numerous examples in different countries (e.g. England, France, Germany, Scandinavia). It can still be traced, in part, as late as the nineteenth century, for instance in the house gardens and the so-called 'warm' and 'cold' fields in Brittany, which were surrounded by uninhabited 'poor' country. An important fact is that these early types of European agriculture were strongly characterised by stock-raising and, above all, the use of the plough; the latter led to more regular field patterns (such as strips or 'block' fields), whereas tropical hoe cultivation, on the other hand, leads to more irregular forms (Sautter, 1962).

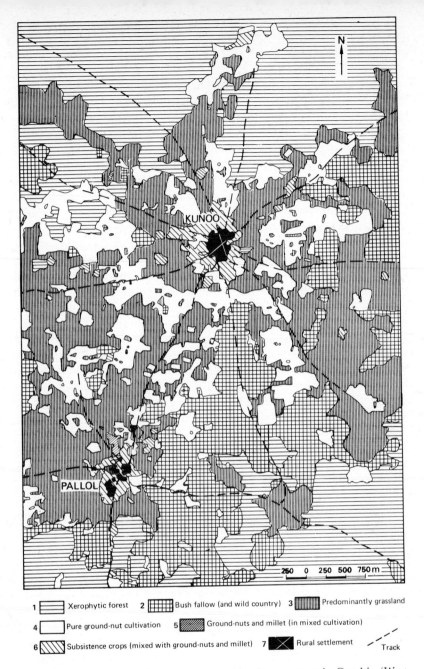

1 [≡≡≡] Xerophytic forest 2 [⊞⊞⊞] Bush fallow (and wild country) 3 [▦▦▦] Predominantly grassland

4 [ ] Pure ground-nut cultivation 5 [▨▨▨] Ground-nuts and millet (in mixed cultivation)

6 [▧▧▧] Subsistence crops (mixed with ground-nuts and millet) 7 [◼⊠] Rural settlement –/–/ Track

FIG. 5.2. This section of map represents a typical land use pattern in Gambia (West Africa) (cf. also Manshard, 1955 and the 1:25 000 Land Use Map, Sheet 8a/III, 1959, 13° 45′–50′N; 14° 50′–15°W). The circular arrangement of land use around the villages is typical of land rotation. The fields wander away from the settlements, the expense of transport between hut and field increases, and at certain distances secondary settlements (offshoots) are founded. The frequency with which changes of settlement location occur depends on a number of factors, e.g., density of population, quality of soil, traffic connections (nearness to roads), type of transport (bicycle).

Around the villages a circle stands out where special attention is paid to crops; these are subsistence and other food crops (6). The fields of peanuts (4) and the grassland areas (3) clearly penetrate into the savannas and the xerophytic forests, which are less influenced by man.

As with the division of the productive area into infields and outfields, so also can it be said of linear settlements with long strips of land at the back (*Langstreifenfluren*) that they are convergent forms which can occur wherever there is a planned collective opening-up of virgin land. The linear settlement, arranged in rows, usually adapts itself to surface conditions (rivers, forest paths, interfluves) (Manshard, 1961b). In these 'shoestring' villages the central road may widen out into a square around which the houses are arranged and which may also contain the market, the parking place for transport and the house of the chief. Even when the whole community of an already existing settlement clears new land in the shape of large blocks, this new land is frequently shared out among the participating families in small strips, reminiscent of other long-strip fields. The survey is carried out from the pathways and takes quality into account in trying to achieve a just distribution of the land. This kind of division mainly takes place in the outfields. The fields alongside the houses are square in shape and are the permanent property of the individual family. On the outfields the allotments are shared out anew in every cultivation period (Sautter, 1962). Such open fields, subdivided into strips, are mainly found in areas where shifting cultivation is no longer feasible because of the high density of population but where there is still sufficient land for fallows of six to eight years to be allowed. Later on, when land becomes scarcer, pure individual property ownership quickly results; then division into strips disappears again.

An example of the distribution of land in strips is the *huza* allotment-system of the Krobo, Shai and other groups in Ghana, which have been described several times (cf. Hill, 1963; Hunter, 1963; Manshard, 1961a, b; 1965a). A *huza* is a piece of land which as a rule is bought as common property by the members of the same Krobo family or clan. Every member of this 'company' gets a piece of land after the big block has been subdivided into strips (*zugbes*), varying in size according to his contribution to the original purchase. The survey of the individual strips causes the greatest difficulties. Rivers or forest paths were first used as guidelines. Such *huza* fields are almost always relief orientated. These strip fields of the Krobo are usually bounded by streams or divides. They can lead eventually to the fragmentation of property. In its structure, this field and settlement form (with isolated strips of land connected to the farmhouse, or with long strips in rows) bears certain similarities to the Central European *Waldhufendorf*. There are parallel cases in other colonial countries too (Brazil, cf. Waibel, 1955), where similar linear settlements with long strips of property have been laid out by colonists.

It can already be seen from this short exposition that it is certainly possible to compare present land use patterns, especially in tropical Africa, with both older and more recent European types. It is obvious that in both cases a division into infields with individual property and outfields with collective property can be recognised as a basic structure. The date at which the outfields became private property can also be proved in many cases. In spite of their resemblance to some old European land use patterns, these African types have definitely developed independently and have not been inspired by external influences (convergence forms). These existing types of tropical agrarian landscape offer modern research a good chance of studying the field patterns of land rotation from actual experience, whereas in Europe they have to be tediously reconstructed from the remnants of the past. As regards future work in this wide field of problems, it seems important that more consideration than hitherto should be paid to the connection between shifting cultivation or land rotation and types of social institution (common property, individual property etc.) (see p. 166).

Irrigated cultivation

With the advent or irrigation man interfered with, changed and usually improved, natural climatic and hydrological factors such as precipitation, underground and surface drainage, and evaporation. Localities could thus be supplied with water which would not be available under natural conditions. Although irrigation systems exist in many climatic zones of the earth, the warm tropical and subtropical belts have been the principal areas of irrigated cultivation for thousands of years.[1]

As an introduction to this chapter, the author would like to try to systematise irrigation techniques by means of a schematic, Fig. 5.3, which illustrates the great variety of technical systems and methods of irrigation that characterise the irrigated landscapes of the earth (especially in the arid climatic belt of the marginal tropics), without taking social and economic phenomena into account.

Compared with the extensive areas under various forms of land rotation these irrigated lands, with very few exceptions, are limited in extent and concentrated round important water supplies. In addition to the ordinary rainfall received, water may be distributed artificially for different purposes. Not only is the water supply of cultivated plants thereby assured, but organic and inorganic matter is also provided which feeds the plants and improves the soil. In contrast to cultivation on rainfed land irrigated cultivation helps to prevent soil erosion. It does, however, present other problems which have to be solved, such as the progressive salination of the soil. Larger and more regular harvests can be achieved with irrigation; but investment is more costly and people have to be prepared to collaborate in sharing water. Because of the need for specialised technical knowledge, a farmer used to methods of rainfed cultivation may take a long time to accustom and adjust himself to the new methods required when the land is irrigated.

Irrigated cultivation must be subject to a precise work plan and the

---

[1] An important English text on this theme is the *World Geography of Irrigation*, by L. Cantor (1967).

NATURAL OR ENVIRONMENTAL PRECONDITIONS	A. IRRIGATION SYSTEM	TYPES OF ORGANISATION

A. IRRIGATION SYSTEM
  I. Sources of water supply
     1. Rivers
     2. Springs, artesian wells
     3. Rainwater
     4. Groundwater
  II. Methods of raising water
     1. Manual methods
        (a) Draw well, e.g. *shaduf* (Arabic), *cigonal* (Spanish)
     2. Worked by animals
        (a) Inclined plane, e.g. *gird* (Arabic)
        (b) Geared wheel, e.g. *noria* (Spanish/Arabic), *nora* (Portuguese), *sakieh* (Arabic)
     3. Automatic devices
        (a) Impounding the flow of a river by a dam or weir
        (b) Kanat
        (c) Water wheels, e.g. *noria* (Spanish/Arabic), *naûra* (Arabic)
        (d) Windmill pumps
     4. Motorpumps
        (a) Diesel pumps
        (b) Electric pumps
        (c) Steam pumps
  III. Water storage
     1. Reservoirs
     2. Ponds
     3. Cisterns
     4. Tanks
     5. Water towers
  IV. Water distribution (supply and drainage)
     1. Irrigation canals
        (a) Main canals
        (b) Secondary canals
        (c) Distributory canals
        (d) Furrows
     2. Drainage canals
        (a) Main drainage
        (b) Secondary drainage
        (c) Collector drains
     3. Levelling or terracing
B. IRRIGATION METHODS
  I. Surface irrigation
     1. Dam or barrage
     2. Dam/barrage and channel irrigation
     3. Channel irrigation
     4. Spraying
  II. Underground irrigation

Irrigated fields e.g., Sahorra (Canary Islands)

Types of irrigated landscape e.g., oases.
In Latin America: *huertas, vegas.*

Left margin (top to bottom): Vegetation, Soil, Relief, Climate

Right margin (top to bottom): Water rights, Water cooperatives

Sources: Geography Department, Giessen, 1967 and attempts at classification by Hirth (1921), Kreeb (1964).

FIG. 5.3. An attempt to systematise irrigation techniques

settlers employed are under strict discipline. As irrigated agriculture can only be successful with strict rules and regulations, many governments refrain from it because it might prove unpopular. Great difficulties arise again and again from the fact that it is usually politicians who decide who should be employed in an irrigation project. Because they court popularity and lack expert knowledge, such politicians often choose the poorest or landless people, or even nomads, with whom rational irrigation cannot be carried out.

Archaeological research has shown that the oldest irrigation cultures can be dated back more than 5 000 years. The early high civilisations of Mesopotamia ('the Land between the Two Rivers') and the Punjab ('the Land of the Five Rivers') were based on the development and mastery of complicated irrigation techniques. It was only by the help of these techniques that agriculture could penetrate into the arid zones of the earth. It is now, however, more or less clear that the importance of irrigation in developing the agriculture and civilisation of mankind has often been overestimated. Both in Mesopotamia and Peru it achieved a certain technical perfection and spatial expansion later than used to be thought, and only attained greater importance after the development of a considerable urban culture (Braidwood, 1962).

PLATE 4. Replanting of rice: Malabar, India. Rainfall is so high that in the swampy coastal lowland enough water can be retained in the fields without irrigation

The planned economy type of organisation which is reflected in the geometrical patterns of irrigation canals puts irrigated cultivation into a special position as an economic type. It can be carried out as both a plough and a hoe culture. On the one hand we find the small irrigation economies of dry belt oases, organised by single villages, and also ponds ('tanks' in Sri Lanka (Ceylon) (see Fig. 5.4) and southern India) or the cultivation of terraces; on the other hand, irrigation is often connected with large-scale projects promoted by the state. The great dams and irrigation systems on the Nile (e.g. Aswan, Gezira), Niger (Office du Niger) or Indus, resemble each other technically in many ways. These large-scale projects only differ in their organisation, being run either by the state or by private enterprise, and in the form of property rights and other social aspects.

In many tropical countries the large-scale irrigation projects, often the byproducts of multipurpose hydroelectric dams, have proved unsatisfactory. In spite of the power generated, the financial investments over many years and complicated systems of centralised organisation were a heavy burden to the countries concerned. Often smaller tubewells and electric pumps have been more successful, particularly in countries that have already benefited from the 'green revolution' and where, because of

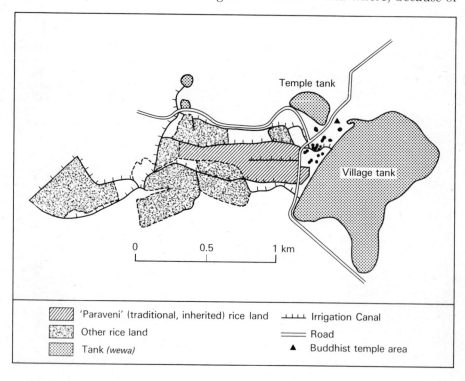

Fig. 5.4. Village in Sri Lanka (Ceylon) with tank irrigation: Pul Eliya (Nuwarakalawiya) (after Sievers, 1964)

PLATE 5. Irrigation tank on the Deccan, India, with inselbergs in the background

favourable farm prices, individual farmers could buy these pumps them-selves. Farmers in the delta areas of Thailand, South Vietnam and Bangla-desh have been able to buy many pumps in recent years in order to lift water from low flowing streams and canals to field levels during the dry season.

So far, in Asia where nine-tenths of the world's rice crop is produced, only one-third is irrigated. Most of the water is rainfed, either by flooding or by trapping the monsoon rains in fields surrounded by low earth dikes called *bunds*. While in Taiwan nearly all rice cultivation is under irrigation, in India and Thailand it is only one-third and one-fifth respectively (Brown 1970, p. 28, 29).

Highsmith (1965, pp. 382–9) put forward a cartographic representation of the expansion of the world's irrigation systems. From this map (the estimates are by Fels, 1965) the large extent of irrigation in Asia is obvious (e.g. China *c*. 740 000 sq km, India 283 000 sq km, Pakistan 105 000 sq km, Indonesia 5 500 sq km. In comparison with these figures, the irrigated areas of Africa appear very small (e.g. Egypt 24 800 sq km, and the Sudan 7 600 sq km).

Two very big irrigation projects are now under construction or in pre-paration in South Asia. In Pakistan the Tarbela Project is foreseen to irrigate an additional 1·8 million hectares of land. In India the Indian Water Grid may prove to be the world's biggest water project ever. This plan calls for the lifting of 25 billion cubic metres of water annually from the Ganges 460 metres up into a reservoir, from where the water is to be distributed across the Deccan plateau through a 3 300 kilometre network of aqueducts, gravity canals, tunnels, natural water courses and reservoirs, thence flowing again into the southern and western rivers of the country. The objective is to increase the irrigated area from at present 8 million hectares to 19 million hectares. With nearly 80 per cent of its 580 million people dependent on agriculture irrigation is by far the predominant user of water in India.

In many large irrigation projects in Africa a great number of crops are grown; the same is also true of small-scale irrigation farming. Cotton is important as a pioneer plant and as a local stimulus to the textile industry. Recently there has been an expansion of artificial spraying, which imitates natural rainfall. This latest technique completes the 'arsenal' of irrigation methods, ranging from the old *shaduf*, the draw-well and treadwheel, the horse-operated geared wheel, the cistern, tank and subterranean channel, to wind motors and the motor pump.

The so-called *kanats* (also called *foggaras*,) are interesting in this context.

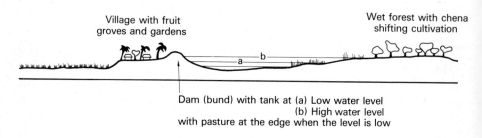

Village with fruit groves and gardens

Wet forest with chena shifting cultivation

Dam (bund) with tank at (a) Low water level
(b) High water level
with pasture at the edge when the level is low

FIG. 5.5. Cultural geographical profile of a Purana tank settlement

The word, which is of Arabic-Persian origin meaning 'pipe' or 'pipes', refers to galleries of groundwater. There are also alternative regional terms. The literature also describes the *kanats* as subterranean water mains (Suter, 1952) or subterranean conduits (Bobek, 1962a). The distribution of kanat irrigation in the Old and New Worlds has been described by Troll (1963b).

Irrigation cultivation in Africa usually implies applying water to land otherwise almost permanently dry. In monsoon Asia, however, one finds alongside this dry-field irrigation a great deal of wet cultivation. This latter is mainly found where there is enough precipitation and groundwater for one harvest; additional irrigation can bring about a remarkable increase in yields, since it permits a second harvest. Rice, the staple food of the population, is the most important crop grown in wet cultivation.

The rice-growing areas of Sri Lanka (Ceylon) may serve as examples of the agrarian structure of this wet rice cultivation. Besides the cultivation which relies simply on the rainfall of the Wet Zone, rice is also irrigated from tanks. These tanks, with their *Strassendörfer* (street villages) alongside them, are so typical of the north and east of the island that one can even speak of a 'tank landscape' and of 'tank settlements' (see Figs. 5.4, 5.5). Cultivation, other than that served by irrigation, is regulated by the monsoon rains. The main period of cultivation is immediately after the summer monsoon; 80 per cent of the ricefields are cultivated at this time (mid-August to the end of November, depending on exposure). Seeding takes place on more than half of the rice land following the winter monsoon (March to April). Two harvests are obtained only in the Wet Zone of the southwest and in some of the eastern tank areas. Cook (1953) even mentions a third harvest which is supposed to be possible in some parts. Compared with other Asiatic countries, yields in Ceylon are low. This is due to old-fashioned methods of cultivation, on the one hand, and to the choice of poor strains of rice on the other. Systematic and intensive manuring is unknown, as are modern ploughing methods and transplanting (preseeding the rice in seed beds and later transferring the seedlings to the fields). The few areas in which transplanting is done are also the areas with the highest yields. The size of the fields varies. Continuous subdivision through inheritance has brought about a strong fragmentation of the land; in the over-crowded southwest about half of the holdings are under 0·2 hectare. In the areas of the east and north newly opened up by tank irrigation, the freshly reclaimed rice land is shared out in parcels of 1·2 hectares.

The rice-farmers' work calendar starts with flooding the fields during the dry season. When the soil is sufficiently soaked it is tilled with a plough resembling the Roman plough. After this, the land is submerged under water again. Six weeks later it is ploughed for the second time and after the water has been drained off it is levelled with a board. Then the pregerminated rice is sown by hand in the fields; in the mountainous Kandy region, transplanting the rice seedlings from the seed beds is only carried out by specially trained rice farmers. The growing season (four to six months) falls within the monsoon season. When the harvest is brought in during the following dry season, the work calendar is completed (Bartz, 1957; Sievers, 1964).

In the Indus region of West Pakistan over-irrigation and salination present such a difficult problem that the very agricultural and economic stability of the country is at stake. The magnitude of this problem has found corresponding attention paid to it in public discussion. One could mention the works by Fowler (1950), Ahmad (1961), Boesch (1962), Blume (1964) and Verstappen (1966).

The complex causes of salination in the cultivated area of the Indus plain have all been more or less thoroughly investigated. They are mainly the following factors: the hot arid climate of the plain; the rise of the groundwater level; water rising and evaporating because of the nearness of the fields to the suction area of the pump holes. It is the combination of these individual factors which may lead to the formation of saline soils.

The excessive application of water to the land is due, to a large extent, to the practice of irrigating all the year round. Particularly serious consequences follow from water losses through canal walls in the alluvial gravels, and from new cultural phenomena such as roads and railway embankments, which considerably obstruct drainage during strong seasonal rainfall. The groundwater level has risen very quickly in the irrigated areas between the rivers (where previously it was at its lowest), reaching the surface in parts and thus rendering the land incultivable.

Salination and waterlogging have thus often appeared as the consequence of large-scale irrigation projects. The irrigation technique applied by the farming population up till then—inundation canals which only filled up and irrigated the land at high water—did not present such threats. It was perennial irrigation of cultivated land which led to the rise in groundwater level and thus to salination and waterlogging. Problems similar to those in

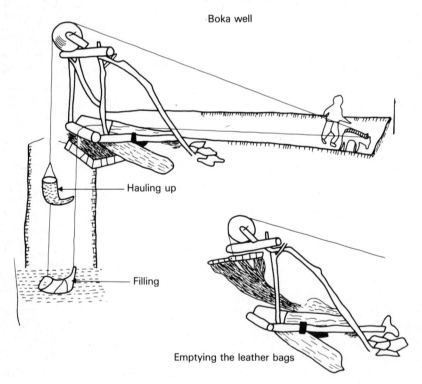

Fig. 5.6. Boka well in Pakistan (after Rahman, 1967)

The draught animal walks away from the well and in doing so hauls the water-bag up over a pulley. The container is then emptied into the irrigation ditch either by a draw-rope or by hand. Then the animal backs up to the well and the bag is filled again when it is submerged.

the Indus plain are experienced in other arid areas where perennial irrigation is common.

Pakistan faces a very difficult problem in the Punjab and the Sind. On the one hand it is necessary to raise the cultivated area and food production a great deal, because of the large population increase. On the other hand the methods used to reclaim new land had led to a considerable loss of cultivable soils, because of waterlogging and salination. The question is how to avoid the negative consequences of perennial irrigation in an arid area

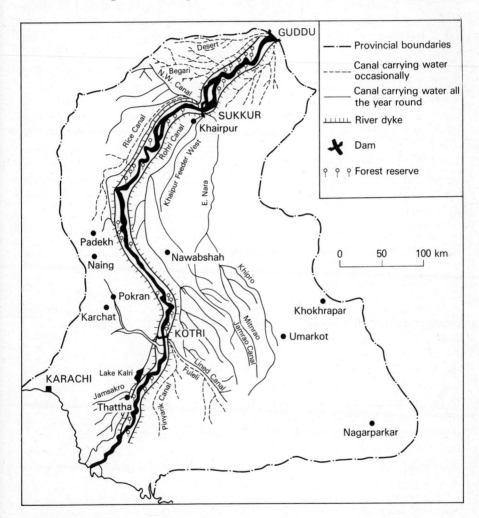

FIG. 5.7. Irrigated areas in the Sind (Pakistan)
The land along the lower course of the Indus receives water by canal irrigation. The canals take off their water from above the dams (only in spring and summer, as far as the seasonal canals are concerned). The length of the canals illustrates the extremely small gradient of the Indus plain (0·3°/oo).

75

such as the Indus plain; there is no doubt that Pakistan's agricultural output can only be increased by some such irrigation technique.

Only as late as the 1950s did irrigation engineers begin to study large-scale measures to combat salination. Previously attention had been concentrated on projects of more restricted local importance. These yielded valuable experience and evidence for later research, however. It became obvious that the only way successfully to overcome the problem is by pumped drainage. The high costs of numerous pumping stations and other plants simply have to be borne. Water pumped from the bottom of wells as deep as 75–90 m can be re-used for irrigation, provided that it is not too salty. In this way water requirements of the plants are covered, while at the same time salination and the upward movement of water are combatted. The simpler and cheaper technique of drainage by open ditches is not feasible because of the low gradient of the land.

One large-scale operation can hardly put things right; indeed, with millions of hectares of salinated and waterlogged land at stake, finance alone can scarcely solve the problem. Agrarian structure and techniques will have to be greatly improved to achieve a real and lasting restoration of the agricultural land in the Indus plain. It will also be necessary to modernise the old Islamic social institutions, with their regressive inheritance laws, adapting them to the new situation.

Plough cultivation and cattle-raising

Transitional farming types in Southeast Asia

Many tropical and subtropical regions do not yet have the type of economy in which cultivation is seasonal (winter or summer, dry or wet season) and linked with the use of the plough and cattle-raising. For plough cultivation, the raising of draught cattle is necessary, and before European colonisation the bases for it were lacking in the New World. In the African tropics nagana and other animal pests made it very difficult, if not impossible. Even at the present day, a central problem remains how to mechanise cultivation in a useful and significant way.[1]

Forms of seasonally cultivating enterprise can be traced from the cold-temperate zone via the subtropics as far as the core tropics, where in some Southeast Asiatic monsoon countries they are overlapped by the zone of

---

[1] Perhaps the right equipment for mechanising tropical agriculture is not simply the tractor but small light implements for use with one or two oxen, which can be applied in many different ways for working the soil, cultivating and harvesting. They are easy to handle and the people only have to learn how to deal with draught animals. Attempts of this kind have been carried out in West Africa for a number of years. Such small implements drawn by oxen have so far been used with greater success than the technically complicated tractor, which also needs more looking after.

permanent cultivation and cattle farming, in which two or more harvests can be gathered each year. These farming systems show many transitions—to intensive irrigation, seasonal shifting cultivation, land rotation and horticulture (see p. 83).

When mixed farming with manuring, crop rotation and foddering had developed in our temperate zones into scientifically guided rational forms, similar practices began to spread in the tropics and brought about many transitions, variants and mixed types, ranging from semipermanent cultivation to the various kinds of pastoral economy.

Especially manifold is the superimposition in Southeast Asia of various land use and farming systems. Plough cultivation and more productive irrigation techniques were probably introduced into Indonesia through intensive cultural exchanges with India; previously, only simple irrigation cultures were known in Indonesia. In the nineteenth century the colonisation of Indonesia became more intense. Plantation cultures were introduced and the economy changed over to export products; this development was concentrated mainly on the island of Java, which was economically most developed. Thus the contrasts between areas with dense and sparse populations among the islands were further intensified. For a long time, on the often fertile, volcanic soils of Java, forms of enterprise have been successful in which different kinds of land use have been practised. In the forefront there is irrigated plough cultivation for rice (*Sawah* system). Beside it, orchards and small vegetable gardens are laid out round the houses. On fields which cannot be irrigated (*tegalans*, dry fields) and on dry 'sawahs', manioc, sweet potatoes, maize, peanuts, soya beans and so on are cultivated, either with the hoe or the plough. Besides small animals (sheep, goats and poultry) water-buffaloes, oxen and cows are often kept. Fish-breeding in sweet- and saltwater lakes raises production on these mainly small but intensively used enterprises. With properties of about 1 hectare per family the land has to be used intensively, and as the density of population is very high there are very few unused acres. Besides wet rice cultivation, mixed horticulture shows great variety. Root crops, vegetables, spices, medicinal herbs and tobacco are planted in the shelter of clumps of trees.

Especially with rice cultivation, it is typical to find these small gardenlike fields in which useful perennial crops are grown. One can also observe a great variety of mixed trees and shrubs (e.g. citrus fruits, mangoes, coconut palms, coffee, pineapples, papaya) which are partly grown for shade. This kind of cultivation, found for instance in Malaysia is not at all restricted to tropical Asia. It is also known in the West Indies and Africa, often even in rather close connection with shifting cultivation and land rotation. In the gardens, a mixed stock of trees often brings about in an artificial way conditions which are similar to those of wet forest; the cycle of nutritive materials also corresponds to that of the forest. As manure is usually applied, because of proximity to the house, soil fertility is maintained longer. Although some tropical experts believe that the future of agriculture lies in an integration of horticulture and mixed tree cultures, the productivity of these groves of trees is low and there is no doubt that plantations of pure stocks would yield more.

With the increase in population density, gardens and tree groves expand even at the expense of the *sawahs*, the result being that root crops become more important in the diet. In recent years *tegalan* dry-farming has frequently become important for cultivating products orientated towards world markets.

Originally, hoe and digging-stick cultivation was common in many agricultural areas of Indonesia. With the development of the plough technique draft animals were introduced: water-buffalo in wetter regions, cattle in drier areas. The *sawah* plough culture of swamp rice (padi or paddy) has expanded from the Chinese and Japanese subtropics far into the Indonesian region. Often several harvests can be brought in on the floodplains of monsoon Southeast and East Asia, if intercrop cultivation is carried out carefully; the techniques, however (e.g. using the sickle instead of the scythe), necessitate a very intensive use of labour.[1]

In the Sunda Islands cultivation originally had a different appearance. First of all, in the hilly and mountainous areas, dry rice was still cultivated by shifting cultivation and land rotation after clearing by fire. However, elements of the 'Java' cultivation system penetrated into many remote regions. Trees were planted round the settlements and as water-buffalo and small animals came to be kept, and fish bred, these formerly different systems became more and more similar. Even in eastern Java, forms transitional between simple land rotation and shifting cultivation (*ladang*) and irrigation can be found. On the dry northern coast and in overpopulated Madura, people adapted themselves to the greater aridity by going in for the dry-farming of maize and the cultivation of small gardens and clumps of trees (e.g. mangoes, citrus fruits). Here, ricefields were not irrigated but depended on rainfall. The Java system with its various main features—irrigated plough cultivation, tree cultures, house gardens and drought intercultivation—has spread for centuries at the cost of older systems (such as the Polynesian horticulture and the *ladang* economy). In many remote areas even nowadays, these old systems can still be recognised as relics. Thus, over a great part of Indonesia (such as Bangka, Billiton, Borneo, Minahassa, Halmahera and many Moluccan islands) the raising of large animals, including cattle and horses, has receded. There, a simple *ladang* system is common, in which rice, maize, roots and spices are planted with pointed digging-sticks on very small fields cleared by fire. Tree and shrub

---

[1] An especially striking variant of rice cultivation is deepwater rice (floating paddy), which is found mainly in the delta areas of Southeast Asia. This type of rice cultivation can be carried out to a depth of 6 m, provided that the water level does not rise more than 10 cm per day. It is more common to find water depths of 2 to 3 m. Rice is also grown in the lagoon areas near coasts, e.g. in India around Madras.

cultures and the keeping of small animals can also be found almost everywhere in these systems.

From the point of view of cultural geography, there is an important contrast between the tropics of Africa and Asia; as is well known, the density of settlement and agricultural production are much higher in many parts of the lowlands of southern and eastern Asia. Many factors can be quoted to explain this striking difference, which becomes especially obvious in a comparison of the great tropical delta areas (such as those of the Niger and the Irrawaddy.

In the deltas of the Irrawaddy, Sittang and Salween, as well as in the southern river and coastal lowland areas of Arakan and Tenasserim in Burma, a pronounced rice monoculture developed. The cultural landscape which was created is reminiscent of that of the deltas of the Mekong, Menam or the Red River. This development in south Burma is relatively young and can be compared with the great pioneering achievements of mankind such as the opening up of the North American prairies or the South American pampa. At the beginning of the nineteenth century almost all Burma's rice, which was already the main element in the diet, was still grown in the northern dry zone. At times, rice exports were forbidden by the kings because the country's own demands could only just be covered. The extensive swamps and lowland areas were untouched, in the same way as large areas in Sumatra and Borneo are today. The opening up of the southern delta area began in the 1840s. The scale of the development can be judged from the remarkable increase in cultivated land (1865: 0·7 m hectares; 1910: 3·1 m hectares; 1940: 4 m hectares; more detailed data can be found on p. 164).

A comparison between the Asiatic and African[1] tropics makes it obvious that one must not underestimate the great importance of irrigation agriculture in tropical Asia, where it has had a long tradition. Swamp rice cultivation, in which the water flow is captured and controlled, and can be used for man even during the dry season, makes it possible to feed a high population on a relatively small area.

Field and settlement types

Stable settlements, often existing for generations, come to play an important part in plough farming. Some of these types of settlement from Ethiopia and Thailand will be sketched as examples.

It is in Ethiopia that one can find forms transitional between an im-

---

[1] In the last few decades measures have been taken to reclaim swamps and grow swamp rice in the West African mangrove and coastal swamp areas of Senegal, Sierra Leone, Gambia and Liberia, in the same ways as in Guyana and Surinam. The main problems have been the clearing of swamp forest; the prevention of penetration by salt water, by blocking off tidal waters; a sufficient inflow of fresh water to guarantee desalination; and hygiene, as malaria and bilharzia can spread with the cultivation of swamps.

proved land rotation and early plough cultivation, developed by the Amhars as they spread from core areas such as Gojjam and settled success-fully in numerous neighbouring regions. An important factor here was cer-tainly the cultural superiority conferred through planned grain cultivation and stock-farming. The introduction of the foot-plough, however, which from the point of view of culture and scientific knowledge was indeed a step forward, had in part negative consequences for the ecology and economy of the Amhar highlands. It resulted in a vast destruction of natural vegeta-tion and especially forest. Slashing and burning and overgrazing finally caused more extensive devastation of the land (Kuls, 1963). Where settle-ments are permanent and the productive land stabilised, different zones of cultivation can be distinguished in Gojjam: garden land, manured land near to the farmsteads, irrigated land and unmanured, extensively culti-vated cropland. All the other land which is not cultivated, be it extensive grasslands, savanna woodlands or isolated mountain forests, is used as pasture. Adjacent to the farmyard there is a small cultivated garden which is hoed by the women, in which spices, cabbage, onions and potatoes are grown. More frequently found than the garden land is a fenced-in field of 10 to 20 ares, which is planted with maize, cabbage, millet and potatoes. This field, as well as the garden land, is manured by the excrement of domestic animals. Often, this 'dungland' may be larger and planted with field crops, which otherwise are only found on the unmanured outfields. In the vicinity of the settlement permanent cultivation is carried out, where-as fallow is practised on the rest of the fields. Only a relatively small propor-tion of the fields can be intensively used through manuring, because the cattle are not stabled all day long. In areas which are completely de-forested, the manure is partly used as fuel. In other parts of Gojjam, instead of 'dungland', which barely comprises 10 per cent of the agricultural area, irrigated land is found near to the farmhouse and intensively cultivated for onions, peppers and tomatoes. In the less densely populated areas of Gojjam a kind of shifting cultivation is still common, leading to a more marked movement of cultivated land and settlements.

Hövermann (1958) mapped typical examples of *Haufendörfer* (nucleated villages) and *Gewannflur* (open fields) in Ethiopia. The fields under per-manent plough cultivation are divided into three units (south, north, east) which are alternately planted with barley, oats, tiff, linseed and lentils. Each field is subdivided into *Gewanne* (communal strips of the same quality throughout) and individual lots, in which rather small strips predominate. The land submitted to such a distribution is called *Medri dessa*—which is why one speaks of a '*dessa*-system'. This form was derived from an older system, the *Medri haut*, in which not all the inhabitants of the village got the same share in the divisions.

Hamlets (with *Blockgemengeflur*, a pattern of fields in which the land owned by each farmer consists of a number of small, widely scattered plots) are mentioned as a further type of settlement and field system. In contrast

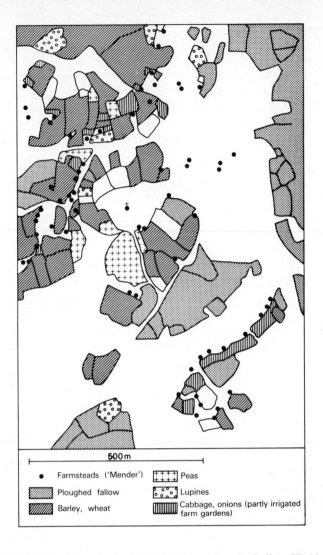

500 m

- Farmsteads ('Mender')
▨ Ploughed fallow
▨ Barley, wheat
⊞ Peas
▦ Lupines
▥ Cabbage, onions (partly irrigated farm gardens)

FIG. 5.8. Conditions of cultivation in the Ṣḥabi 'Gemarkung' (Gojjam Highland) (after Kuls, 1963). The 'Gemarkung' is the utilised area of a village, including fields, pasture, exploited forest and built-up area.

The conditions of cultivation (see also Fig. 4.5) are defined by a system of two small fenced-in fields (*Zelgen*). In the eastern part of the district is the fallow field, which is ploughed once or twice, whereas the cultivated area to the west and north is planted with grain (barley or wheat) and lupines. Other available plots in this area are usually sown with lupines as well, without the soil being broken up for the purpose. Near to the individual farmsteads is the manured land, on which peas and barley can be planted. To improve the soil, cattle dung is used. Immediately next to the farmsteads, for the most part, there are partly-irrigated kitchen gardens with onions, cabbage and other vegetables. Kitchen gardens – and usually the manured land as well, are not included in the *Zelgen* rotation; permanent cultivation takes place here. The rotation is carried out in a four to six years' cycle. The farmsteads are grass-covered round huts in which animals are kept as well. Separate cowsheds, barns or granaries are very rare. The land northeast of the church has been kept free for a planned administrative centre.

to the *Gewannflur*, the fields here are subdivided into individual blocks. Land tenure conditions are different, too. The *dessa* system is not applied. Each farmer owns strictly delimited fields and can dispose of them as he wishes, i.e. sell them or bequeath them. In cases of inheritance, division of the property is usual. Land use corresponds with these conditions of individual ownership; there is no order of crops, each farmer growing what he thinks best.

In the same way as Hövermann in Ethiopia, Sautter (1962) has described phenomena in West Africa which have parallels in our own medieval and premedieval field and settlement types. The question arises whether it is possible to compile a general comparative systematisation of world settlement types which would express the social and economic evolution of human groups. Certain analogies of field and settlement form can be seen even in the structure of seigneurial organisations. In the north Ethiopian area where the forms of field and settlement already described are common, feudalism was pushed back by communities of free peasants. In other areas colonised by the Amhars the feudal structure still distinctly exists. Last but not least, it is reflected in the extensive properties of the emperor, the feudal landlords and the monasteries. There is a need for further study of the striking parallels between medieval settlement and field types in Europe and recent forms in the tropics.

In the tropical highlands of East Africa success with mixed farming came with European settlers, on the basis of plough cultivation and stock-raising. In the Kenya Highlands, which constitute about 20 per cent of the area of Kenya and form one of the most extensive highland regions of East Africa, various ways of applying European agricultural methods could be observed, depending on climatic and economic conditions. In the dry areas of the East African Rift Valley large cattle-breeding farms exist on which European and African cattle are raised. The large plantations which have been developed on the highlands, growing coffee, sisal and tea, were not so typical as the European farms which have existed for several decades and on which grain was cultivated and cattle raised in mixed enterprises, 200 to 600 hectares in size.

At the outset, it was common practice to keep many fields under permanent cultivation, but as a result of progressive soil exhaustion and low yields (by European standards) crop rotations were later introduced, with fodder plants, millet and sunflowers. So too were the very important pyrethum, a plant which is used for the production of an insecticide, and the tannic acacia ('wattle'), of which the extract is used in leather manufacture.

As a further example of the structure of rural settlements based on tropical plough cultivation one may cite settlement conditions in Thailand. Here too one finds, side by side, very different cultural and economic types; from tribes of hunters and gatherers to plough farmers. The greatest portion of the population are Thai peoples, whose economy is based on the plough and who live in the fertile plains and grow rice in permanent cultivation.

In founding a settlement, proximity to water is generally preferred. If a village is situated on a river the rows of houses follow its banks, their entrances facing the river on the principle that it is the main traffic artery (riverbank 'shoestring' village type). A special characteristic of Siamese plough-farming villages are the tree cultures which surround almost every settlement. *Strassendörfer* (street villages) also occur frequently. In the cattle-raising areas of the north *Runddörfer* (round villages) prevail. These settlements are surrounded by thick hedges of bamboo. Each house has a small granary. As the terrain is unfavourable, rice cultivation plays a minor part.

Special forms of cultivation: horticulture and terracing

Horticulture, carried out in the vicinity of the dwelling house, is probably one of the oldest forms of agriculture. There is no doubt that it has many sources of origin. In the historical evolution of agricultural development, this first step in more or less planned planting of crops was evidently so difficult that some groups of people have not undertaken it even yet, but vegetable production in gardens alongside the houses has remained one of the most important sources of food supply in many parts of the tropics up to the present day. Plant growth in these gardens is encouraged by the use of household waste and human and animal excrement, and weeds and vermin are more easily kept at bay by intense working. These micro-types of agriculture can take the form of seasonal, permanent or irrigated horticulture. Often, horticulture has been retained for more specialised production alongside other systems of agriculture and types of economy (hunting, fishing). Cultivation alongside the dwellings is especially suitable for valuable and sensitive plants whose purpose is to improve an otherwise monotonous basic diet.

As horticulture is typically associated with a close settlement pattern, it is to be found mainly in those regions of the world with the greatest densities of population. Terra (1953, 1957) expressed the idea that horticulture was originally the preserve of women, whereas the men of the community were responsible for cultivating the adjacent fields. It is characteristic of rural areas in the tropics that one often finds, in the heart of extensive forests, intensive horticulture and more extensive forms of shifting cultivation side by side.

In Polynesia and Indonesia very old-fashioned kinds of horticulture are still partly in use. Besides fishing and hunting or food-gathering, tubers such as yams and taro, or bananas, are planted with the aid of the digging-stick or even the hoe. These gardens, next to the dwellings, are occasionally irrigated. The breeding of small animals (goats, sheep and pigs) complements this type of economy. The make-up of these cultures has been changed by foreign influences, in many cases. New plants such as millet, rice and later maize have been adopted, so that a more variegated horticulture has developed, characterised above all by a great range of perishable fruits, vegetables and other useful plants. Very often, this horticulture is not at all a labour- and capital-intensive type of economy, but rather a preliminary stage of the *ladang* economy with land rotation.

In some areas of Africa, too, one finds manured fields being cultivated near houses (kitchen gardens). Groves of trees can often be seen between the gardens around the settlements, especially bananas, for instance in Rwanda, Burundi and Uganda. Frequently, useful plants creep up the walls of the huts, along the bamboo fences or prickly hedges which divide the holdings. Kitchen garden, farmhouse field and infield merge into each other here, as the distinctions between them are not clearcut.

These native gardens, however, cannot be considered at the same level as European horticultural enterprises, which have generally specialised in certain crops. In some areas of southern and eastern Asia, and also in the Andean countries, they rather take the form of hoe cultivation somewhat intensified by manuring and irrigation.

In contrast to the primitive forms described above, horticulture has been developed to a higher level, particularly in those areas of Southeast Asia with a high density of population, being perfected by Chinese gardeners (for instance, in southern China and Malaya). Where the holdings are small cultivation is extremely intensive. All the work is carried out manually by the family. Large animals are not kept. The keeping of pigs, poultry and ducks, and fishing, often helps to complement the diet and contribute to the well-known variety of Chinese food. Unfavourable soils are systematically brought under cultivation by skilful terracing, manuring and irrigation. Thus unusually high yields can be achieved through intercropping, the cultivation of alternate rows of, for instance, rice, tubers and vegetables, and cultivation at various vertical levels. The East Asiatic vegetable grower who has specialised in intensive market-orientated crops can also do well in other countries, as for example, in the Chinese market gardens around Singapore (Ng Kay Fong, Tan Chee Lian, Wikkramatileke, 1966). Interesting examples of a horticulture successfully transplanted from Asia to South America can be observed in the alluvial soils of Amazonia, where Japanese colonists have succeeded in creating productive vegetable gardens in the rainforest. On poorer soils the Japanese settlers grow black pepper, which reached Brazil via the Botanical Gardens in Singapore; its cultivation literally led to a boom in the area east of Belém, Brazil (Pfeifer, 1962, p. 186). Even so, the output achieved with this extravagant use of manpower is still low compared with the yields won by mechanised methods of cultivation in North America or Europe.

Horticulturists' settlements usually consist of individual houses, often surrounded by shade trees and their individual small kitchen gardens. Where tubers are grown, there is a fallow period after some years. Banana gardens, on the other hand, are kept in cultivation for many years. The individual cycle of cultivation depends on the quality of the soil.

Settlements and their fields are often composed, therefore, of dispersed houses and gardens, arranged along the course of a river or in irregular groups, very rarely around a compact village square.

Terraces and ridged fields are the most important agricultural land forms and each has a number of subforms.

The worldwide extent of terrace cultivation, with and without artificial irrigation, indicates that it can be related to predominantly natural environmental conditions. In numerous places, from the wine terraces of western Europe to the tea gardens of Ceylon, the technique is employed by entirely different cultures in quite distinct habitats. This is not at all to discount the processes of cultural transference and diffusion explaining the wide spread of this particular technique of cultivation.

Terrace cultivation demands of a culture a certain level of technical attainment. Terracing changes the natural relief and water economy, influencing soil formation and constituting the prerequisite for a special type of agrarian landscape. The large irrigation systems of the wide river valleys of the Nile, Euphrates, Tigris, Indus, Mekong and Hwang Ho called for greater political centralisation and the employment of masses of workers. In contrast, the early development of terraced slope cultivation frequently takes place in peripheral regions. It is not initiated by large enterprises but by small groups depending on mutual cooperation. Possible areas of origin are the eastern Mediterranean, the semi-arid regions of the Near East and, for wet cultivation, China.

Terrace cultivation for wet crops was developed in many parts of the tropics, even under unfavourable terrain conditions, by groups of people who already possessed differentiated techniques of cultivation. At the same time, soil erosion could largely be halted by this method. In arid zones such as the Yemen and Peru, terrace systems are often connected with mountain streams whose waters are conducted to the fields by canals running parallel with the slopes. On North Luzon (Philippines) the old terrace systems for rice growing were constructed with extreme skill; irrigation cultivation here has left a profound mark on the landscape. This applies especially to the terrace cultures of some of the mountain peoples of North Luzon. These terraces do not fit into the framework of the irrigation cultures mentioned so far; in this case we are dealing with digging-stick agriculture on artificially irrigated terraced fields. The mountain slopes are often terraced from the valley bottoms right up to the ridges. The retaining walls, which are up to 15 m high, run parallel to the contour lines and follow even the smallest indentation. The height of the walls and the width of the individual fields depends on the angle of incidence of the slopes. The field paths follow the narrow, partly stone-capped crowns of the embankments.

In the north of Luzon, the main cultivation of rice takes place before the winter monsoon rains; the exact date for the individual cultivated areas can vary as much as one or two months, depending on climatic and topographical differences. Under the influence of the rice farmers in the coastal plains it became customary to sow the fields a second time, before the summer monsoon rains. Formerly, batate was commonly grown on the terraced fields during the summer; the heavy increase in population in the narrow valleys, however, necessitated rice cultivation as well, during the second half of the working year. Cultivation methods are well developed and adapted to the natural conditions; the difficult terrain makes the use of the plough and water-buffalo sometimes impossible. Transplanting and manuring (both animal and vegetable) are practised everywhere.

The fields are mostly the private property of individual families. Where the property is substantially larger it is usually linked with numerous social duties such as supplying food to the villagers in bad times, holding religious celebrations and financing public festivities.

The work calendar for the winter monsoon period begins about October with the mending of dams and irrigation canals and the turning over of the flooded fields. For this work a digging-stick is used, or in some isolated areas a wooden spade. The water is then drained off in order to bring manure on to the fields. Transplanting begins from the end of November, from the seed beds to the newly flooded fields. The harvest, in the months of June and July, is a communal task since the climatically favoured fields in the valley bottoms are the first to be ready for harvesting. This type of terrace cultivation with artificial irrigation generally demands maximum cooperation among the villagers.

The site and morphology of the settlements vary a lot and often depend on the tribe to which the settlers belong. Whereas the Ifugao build small, hamlet-like villages on piles on steep interfluves or on narrow level shelves on the mountain slopes, the Bontoc lay out their large villages of up to 1 500 inhabitants down below on the broad river terraces or alluvial cones.

Terraces are also common in Java, North Sumatra (Batak), Kashmir, Assam and Ceylon. It is worth mentioning the remains of Maya terraces in the New World. Where soils are thin it is common to spread loose garden mould, rich in humus. These artificial deposits, called *montones* and *conucus*, were of some importance in the West Indies and eastern South America.[1] In some cases—as in Mexico, Kashmir and along the Chinese rivers—this artificial spreading of soil was even carried out on anchored rafts.

Types of terrace cultivation were known in tropical Africa, too, in precolonial times. Thus, when the European colonial administrations strongly promoted terracing as a measure against dangerous soil erosion, it already had native antecedents in Kigezi, southern Uganda (Manshard, 1965b), Rwanda, Burundi and especially northeast Africa (Ethiopia). In many parts of West Africa (Nigeria, Cameroons), in East Africa (Kenya, Tanzania) and in south central Africa (Rhodesia) the decay of the old forms of terrace cultivation was linked with migration out of the inaccessible highland

---

[1] Cf. also the interesting relics of ancient ridged fields, partly pre-Columbian Indian in origin, in Colombia and other parts of South America (Parsons, Bowen, 1966).

regions where the people had sought shelter from slave hunters and military
forays. When greater safety and freedom of movement were restored under

Plate 6. Rice terraces in the Chenab valley, Kashmir

the protection of the colonial powers, many upland settlements were abandoned. The skilfully built ancient terraces decayed or diminished in extent. It was very much later that terracing was started again as part of measures against soil destruction.

Examples of this process can be found especially in Nigeria and north Cameroons. In central Nigeria the animistic 'pagans' retreated into the pathless mountain areas away from the better-equipped groups of horsemen such as the Fulbe and Kanuri. On the steep slopes of the Jos uplands (Bauchi Plateau) or Mandara uplands (Cameroons) skilful terrace systems were constructed, often fortified by dry stone walls. On these terraces millet and beans were grown. Animal dung was spread to improve the soils and useful trees were planted to secure a supply of firewood and timber. In the course of later downhill migrations to the neighbouring lowlands, during the colonial period, these methods of cultivation were given up and a simple land rotation in the form of shifting cultivation and bush fallow was re-adopted.

In the Yoruba area of West Nigeria similar hill settlements (Gleave, 1963, 1966) were also given up, although terracing had not been carried out there. The inhabitants of small fortified settlements in the Nsukka-Okigwi cuesta country of Eastern Nigeria still possess some terraced fields nowadays; they are not familiar, however, with artificial irrigation, i.e. conveying additional water to their land (Floyd, 1965).

Permanent cultivation: plantations

Permanent cultures are those involving the cultivation of perennial crops, whose products are used up directly, or sold, or which are made marketable through industrial processing. First in this category come tree and shrub crops such as coconut and oil palms, hevea, cinchona, citrus trees, cocoa trees, cloves, coffee and tea bushes. Sisal and sugar cane, which also grow for several years, also belong to this category. A large proportion of these permanent crops is not grown on large plantations but on half-commercialised farms or small-scale peasant holdings. Market orientated crops such as cocoa in West Africa or coffee in East Africa and South America may be accompanied by the growing of subsistence crops. In many cases, both are grown in the same fields. This has the advantage that the ground has a regular plant cover; however, it hampers the use of scientific methods of cultivation (manuring, treatment with pesticides, etc.) which is easier when monocultures are practised.

As suggested in earlier chapters, it is common to find field or garden cultivation and permanent cultures being carried on in the same area. As independent cultures they can be found both on plantations and market orientated peasant holdings in the tropics, and in the groves (e.g., olives, agrumes) and orchards of subtropical and temperate latitudes. Although output from these cultures is mostly market orientated, they are also important as subsistence crops (e.g. the date palm groves of oases). Permanent cultures are most common in the regions whose economies are complementary to those of industrial countries, and to a large extent they cover the demands of the latter for luxuries (e.g. coffee, tea, cocoa), oil (e.g. oil palms) and industrial raw materials (e.g. rubber, sisal).

Plantations and estates—The problem of definition

'The plantation is a large-scale agricultural and industrial enterprise, pro-

PLATE 7. The Kola belt in western Nigeria, three km south of Shagamu (about 6° 47'N, 3° 40'E)

Vertical air photo (June 1965); height of flight: 2 000 m.

The photo shows a section of the wet forest belt of southern Nigeria, an area with very little differentiation in relief. The natural vegetation would be evergreen forest, of which a few degenerate remains can still be recognised in the wet lowlands in the northwest and northeast parts of the photograph. The most important economic plants grown on the farms are cola trees (*Cola acumunata*, *Cola nitida*, among others), with their rounded crowns. Cola plantations of different ages can be recognised. The fields, which are worked in a land rotation system (with bush fallow), are planted with cassava (manioc), yams and maize as intermediate crops. In addition, bananas (plantains) grow in the lowlands to the northeast. There are only a few farmsteads in this section of photograph.

PLATE 8. Air photo of plantation cultivation in eastern Nigeria, about 15 km north of Calabar

Vertical air photo; height of flight: 930 m (February 1965). The light and dark contrast is caused by cloud cover (infrared photo).

The northern half of the picture shows part of an old oil palm plantation. A lot of the high trees have already died. The picture presented is thus of a disordered, thin stock.

As the rentability of oil palm plantations was partly hit by government measures, *Hevea brasiliensis* has succeeded it on this plantation, for rubber production. Young rubber trees, not yet a year old, can already be recognised on the southern section of the photograph.

In order to avoid erosion on the bare soil, terraces have been constructed on the steep slopes to the southeast and leguminous plants, which enrich the soil with nitrogen, have been planted as a 'cover crop'. These creeping plants also cover the trunks of felled oil palms, which lie about everywhere. A free space, however, is left around every rubber tree.

The main road of the plantation runs north–south. From it, small drainage channels lead to the plantations. Two dwellings for important employees lie on the main road. The modern workers' homes ('labour lines'), each with a small house at the back for washing and cooking, lie on a minor road in the west of the photo.

ducing high value vegetable products, usually under the management of Europeans, and involving great investment in labour and capital equipment.' This classic definition by Waibel (1933) subsequently proved inadequate and needed amplifying. Since industrial activity is not an absolutely necessary characteristic of plantations, Gerling (1954) describes this kind of enterprise as an economic type which allows of both extensive and intensive methods in the production of tropical and subtropical raw materials; it has to be a large-scale enterprise and it may have industrial

PLATE 9. Sugar plantation on Kafue Pilot Polder (Zambia). Water for irrigation is pumped from the canals in the foreground

features. A plantation having all the characteristics of a capitalist industrial enterprise was and still is mainly designed for the production of a single crop (monoculture). The economic and social premises for the plantation economy have changed, however, in the course of time. Whereas formerly production was mainly concerned with staple products which could be stored, and the question of labour was easily solved by a single purchase of slaves, for the highly mechanised and specialised plantation economy of today capital, labour and also land tenure conditions have become major problems. Sales are generally made on the world market, and because of this dependence there is a special sensitivity to market fluctuations. On the other hand, it is often possible to make considerable profits. As it was rarely lucrative, at least in the beginning, to supply indigenous populations with agricultural products, the plantations, which were predominantly managed by Europeans, had to produce large quantities of products for the world market, maintaining as stable a quality as possible. Great opportunities opened up in particular for the cultivation of industrial raw materials such as manila hemp, sisal, cotton and various kinds of rubber. Only in recent decades have these plantation enterprises, after having been nationalised or reorganised, started supplying the inland markets of tropical countries to an increasing extent (e.g. sugar plantations in Indonesia and Venezuela).

Waibel's and Gerling's definitions of a plantation are no longer so useful as they were before the Second World War. In regions outside the tropics types of agrarian enterprise can be found which compare structurally with traditional tropical plantations but surpass them in their degree of special-isation, mechanisation and rationalisation. This applies to the United States (California) as well as to the Soviet Union (Turkestan) and western Europe. The formerly predominant monocultures are decreasing and sales are no longer aimed at overseas markets. Even the entrepreneurs, nowadays, frequently belong to the same nation or race as the workers, who are them-selves better trained than in the past. Thus the classification once proposed by Gerling (1954) according to type of production, which gave seventeen types of plantation, would now have to be amplified as far as social and economic features are concerned. If one wishes to distinguish the term 'plantation' with respect to other types of agricultural enterprise, it is usual to do this with the help of three criteria:

1. The size of the cultivated area must be over a certain minimum acreage (about 50 to 100 hectares).
2. The main purpose of the enterprise must be production for inland or export markets.
3. Extensive technical-industrial installations must exist, which are neces-sary for the marketing and sale of the products.

A further criterion, still important nowadays, is the differentiation be-tween enterprises run either by independent owners ('planter proprietors') or by the employees of large plantation companies. The independent planter, who is at the same time his own manager, is characteristic of many smaller plantations, e.g. in East Africa. In contrast, the large plantation companies own several big plantations which may be in completely different countries.[1] An interim definition of a plantation (based on the factors mentioned above) might be as follows: a plantation is a large-scale agricultural enterprise which mainly provides products for the market (domestic or international). A special characteristic of this type of enter-prise is the existence on it of processing facilities. This makes for a greater intensity of labour and capital. The plantation is not tied to a particular region, but it is found mainly in the subtropics and tropics.

In contrast to peasant holdings, plantations or estates usually have a tendency to monoculture. In the last few decades, however, numerous

---

[1] Some agricultural geographers (Dietzel, 1938) regard the *Pflanzung* as a smaller version of the plantation. They both have the same economic objective, but in the former case the planter disposes of less capital and his enterprise is smaller and less mechanised. According to this not very happy definition of the two terms, the difference between *Pflanzung* and *Plantage* (plantation) is one of scale rather than structure. Both terms are often used differently in different regions and the lines between them completely erased; for instance, in East Africa the term *Pflanzung* is used for large-scale enterprises while in the West Indies, on the other hand, the terms 'plantation' or 'estate' are applied to the same thing.

monocultivating enterprises have been changing over to a greater variety of products. In such cases, the main crop is complemented by additional types of cultivation. Thus, for instance, coffee and papaya (East Africa), oil palm and cocoa (Zaire) and coffee, mangoes, avocadoes and bananas (El Salvador) are planted either in alternating rows or in separate fields. Where suitable, this extensive permanent cultivation is supplemented by stock-raising (dairy-cattle breeding on the coffee plantations of Brazil or on the coconut plantations of Ceylon (Sri Lanka) or East Africa).

The susceptibility of monocultures to economic crises may be compensated by an ability to react more quickly to changes in market conditions by making the necessary changes in production, unlike 'native cultivation'. Changes made within the enterprise itself, in order to adapt to new market conditions, are not always sufficient. Monocultures often jeopardise the ability of the soil to regenerate itself. In spite of all the phenomena of soil exhaustion and exploitation, which have been known since the early days of the plantation economy, the greatest progress in the sphere of tropical agriculture has been achieved within the context of plantation cultivation.

There is no doubt that the term 'Plantation' (including its manifold translations and variations such as, for instance, *Plantage, plantación, plantacão, hacienda, fazenda,* among others) can still only be used today as a comprehensive general term. A modern definition would have to indicate the flexible and often strongly speculative character of this type of enterprise, as well as stressing its agricultural–industrial features, its intensive use of labour and capital, and the special traits of its organisation.

When Gregor (1972, p. 722) speaks of the quasiplantation as a conceptual model, he thinks of an intermediate type between the lower intensity livestock ranching and the greater intensity crop plantation. This type,

Cultivated products	Luxuries and/or raw foodstuffs Sugar, coffee, tea, cocoa, bananas, etc. Producing for and dependent on world markets	Industrial raw materials Tung, sisal, rubber, etc. (partly monocultures) Endangered by technical progress (e.g. synthetics)
Property types	Private property	Company property High capital investment	State property
Social structure	In charge of the plantation: manager, administrator and the like. Supervisors: foremen, mandars (India), tyndals, kangani, etc. Workers: slaves, contract labour, seasonal labour, coolies, *peones*, *colonos*, and the like. Workers' settlements: compound, labour line, coolie line, senzala (São Tomé), batey (Greater Antilles), bangsal (Malaya), and the like.	
Technical installations	Processing station: e.g. *ingenio, centrale* (sugar cane), *beneficio, secadero* (coffee). Means of transport and machines: e.g. tractor, light railway, sugar mill, etc.	

Fig. 5.9. Some formal elements of the plantation system

which is particularly prominent in California and Texas, may also have interesting possibilities of application in the tropics.

Historical development of the plantation

It is an interesting historic fact that a kind of plantation system was already practised by the Romans in North Africa; wine and oil were produced on their large *latifundia*. According to Ritter (1822) and Waibel (1933) tropical plantation cultivation is derived not only from this root but also from Persian practice in Chusistan, where sugar cane was cultivated. Cane sugar production spread from there into the Mediterranean; at the end of the fifteenth century it was introduced into the Canary Islands, Madeira and for the first time into the core tropics, in the island of São Tomé in the Gulf of Guinea. In the middle of the sixteenth century it spread to the West Indies and Brazil. With the introduction of Negro slaves, plantations in the South American tropics experienced a rapid economic boom, and sugar and coffee plantations developed. Based on experiences in Central America, plantation cultivation later returned to the Old World. In the middle of the eighteenth century sugar plantations were founded on the islands of Mauritius and Réunion in the Indian Ocean, and at the beginning of the nineteenth century coffee and cocoa plantations revived on São Tomé. Up to the end of the nineteenth century a limited expansion of plantations can be observed in Africa, especially in the Cameroons (bananas), Tanzania (sisal) and the Portuguese colonies. Plantation cultivation reached its greatest importance in Asia in the course of the nineteenth century. Here it was not originally sugar production which was foremost but the cultivation of tea, coffee, Peruvian bark (cinchona), tobacco (after 1880 in Sumatra) and rubber (after about 1900 in Malaya). The economic lead of the American tropics in the seventeenth and eighteenth centuries was mainly based on the favourable geographical position of the New World as regards its connections with West Africa and Europe at the time of the well-known triangular sailing-ship trade. After the freeing of the slaves, American plantation owners had great difficulties in procuring labour. At the time it was even hired in Asia, to replace the Negroes (for instance, in Guyana). The centre of plantation cultivation shifted to the Asiatic tropics, as it was easier there to cover labour demand. However, a less favourable position than before, in distance from European markets, had to be supported. It was only when the Suez Canal was opened (1869) that shipping routes were shortened. The Asiatic plantations reached their peak at the end of the nineteenth century and the beginning of the twentieth. In the course of their development great labour migrations took place in south and east Asia (e.g. Tamils from southern India to Ceylon and Malaya). There has also taken place a considerable transfer of Indians to Fiji, where they continue to grow sugar and now actually outnumber the native Melanesians (Indians 241 000, Fijians 202 000 in 1966).

Plantation systems developed above all in areas which were thinly populated and in which more extensive acreages could be cleared and cultivated (note, for example, the contrast between Sumatra and Java).

The reason for the rapid growth of the plantation system is that in Europe, since the early Middle Ages, an economy based on payment in kind was rapidly replaced by a monetary economy. Especially in the period of mercantilism, European states found in their colonial plantations new sources of income and high profits. It was the liberal economic attitudes of the nineteenth century which brought about the greatest changes in the system, whereby financially well-to-do coloured people increasingly began to appear as plantation owners. Already in the Middle Ages Arabs had plantations in the Indian Ocean (e.g. clove gardens in Zanzibar). There are numerous examples of coloured plantation owners in modern times, such as the Chinese in Malaya, mulattoes and mestizos in Latin America, and Afro-Americans in Liberia.

In Brazil, the strong ties between monocultivating large estates and the international market led to a regular cycle of land use. During the first phase, it was mainly sugar cane which was grown on extensive plantations and which formed the basis of the economy. Cultivation was limited to the well opened up coastal strips. Then, at the beginning of the nineteenth century, coffee came to Brazil. The prospect of high and rapid gains made a majority of plantation owners give up sugar cultivation and grow coffee as a monoculture instead. Frequently the parties involved were speculative large-scale enterprises from the coastal towns. The consequence was enormous coffee plantations in the newly reclaimed forest areas near the coast. In view of the general lack of fertiliser or manure, these freshly cleared woodlands with their fertile *terra roxa* were the favoured sites for coffee. After a few decades, however, the soil was leached and its poor capacity for regeneration made it necessary to find more virgin land, either by seeking out fresh forests to clear or by changing over to stock farming. Thus, a final stage in the cycle of land utilisation, as outlined by James (1953), was reached.

The ecological consequences of these monocultures were grave, especially as no shade trees were planted between the coffee shrubs, so that nothing prevented exogenetic forces from operating. Deeply degraded soils need years of rest before they can undergo intensive cultivation again, and then not for monocultures but for a crop rotation of rice, beans and maize (and dairy farming, too, in part). Strong erosion has cut deep fissures into the surface, especially where the rows of coffee shrubs were laid out vertically over the slopes (Pfeifer, 1962). In specially serious cases of soil destruction, a secondary scrub vegetation appeared which only permitted extensive stock farming. The appearance of extensive areas of the southern Campo Cerrado is undoubtably influenced by manmade forces.

Important centres of cultivation during Brazil's coffee period were the areas around São Paulo and the gently westward-sloping plateau between Campinas and the Río Paraná. The demand for new forest soils for the laying out of plantations, and the exhaustion which began after a few years, brought about the creation of a coffee 'frontier' which penetrated further and further to the north and west into virgin areas (Platt, 1935; James, 1953). The best account of the 'Coffee trail and pioneer fringes' is given by França (1956), together with a series of maps showing the extent of the coffee plantations at various dates between 1836 and 1950 (cf Fig. 5.10). Fertile river valleys (like that of the Río Paraíba) and railway lines were followed. As a matter of fact, these were indispensable for the creation of plantations, since the *fazendeiro* depended on good connections between his property and the export port. Gutersohn (1940) shows the importance of a coffee port to its hinterland, quoting as an example Santos, which is the port of transhipment for the whole state of São Paulo.

Up to 1888 the proprietors of the large estates covered their demand for labour with slaves, which came from the African coast opposite. After the abolition of slavery they changed over to the *parcería* system: European immigrants, who came into the country in great numbers, were employed as agricultural labourers and as share tenants. At peak periods the owners also employed migratory workers. These facts are reflected in the pattern of settlements: the buildings for the owner or his manager were erected on higher ground which commanded open views, whereas the share tenants and plantation workers lived in rows of houses along the roadsides.

For the first four years after such plantations have been laid out, and while the coffee shrubs do not yet yield anything, mountain or upland rice is grown to bridge the gap. It is sown between the young coffee seedlings and the tree trunks which are still lying about after burning; it also serves to meet the plantations' food demands in later years. This cannot, however, be called a genuine mixed agriculture, since the growing of rice, and also beans, is confined to those parts of the plantation near the house.

Since the plantations grow their own subsistence crops and coffee for the international market simultaneously, to a large extent they are economically independent of the town. A functional relationship in the sense of town-*umland* relations, or Thünen's Rings, cannot come about. The only thing that matters to the plantation is rapid access to the world market, i.e. a railway or well-built road to the export harbour.

The extent to which a cultivation system like this depends on the world market, and is susceptible to crisis, was shown in the sudden end of the coffee boom in 1930. As a result of this experience many plantations changed over to mixed cultivation, mainly of cotton, cocoa, rice or maize, the Paraíba valley occupying a special position with its dairy farms and citrus plantations—the 'inner Thünen's Ring' of Río de Janeiro.

The discussion whether the plantation system is good or bad for developing countries in the tropics poses questions which go far beyond the field of agricultural geography. Economically there is a lot to be said for plantations:

1. The application of modern farming methods is made possible, with tested crop sequences and better seed control; the spreading of insecticides and the combatting of pests, and thus the achievement of higher yields, which often exceed the output of native agriculture many times over.
2. Their rational, almost industrial, style of cultivation reduces costs. Lines of communication can be built more easily. Since the plantations control processing industries, better quality products can be produced.
3. Advocates of the plantation system point to the successes which have been achieved by training local workers, building schools and hospitals, and distributing food rations which provide better nourishment for the labour force.

On the other hand, the disadvantages must not be overlooked. The wage labourers are uprooted from their old ways of life. Male workers are frequently separated from their families. Difficulties and quarrels with local proprietors over questions of land purchase and leasing are common. The strongest argument used by opponents of the plantation system nowadays is a political one. The nationalists of many developing countries would like to see plantations abolished as 'relics of colonial times', and they wish to hand over the land as smallholdings to peasants. Words such as 'exploitation' are often used superficially, without considering the positive side of the plantation system. One way out has been to create closer partnerships between the governments of tropical countries and overseas investment companies. In summary: it may be asserted that for further agricultural development in the tropics, large and middle-sized enterprises which can serve as growth points for neighbouring farms are likely to be more important than large-scale pilot estates whose value as demonstration schemes has in most cases proved poor.

Peasant farming

In recent decades native peasant enterprises[1] and small plantations have developed more strongly alongside the large-scale plantations, which were world-market orientated and labour and capital intensive from the outset.

An important prerequisite for the rapid development of world-market orientated permanent peasant agriculture was the stabilisation of property and land tenure conditions (Manshard, 1965). Originally, many groups had laws which forbade the sale of land, the individual being unable to dispose of his land as he wished. In some tribes, landed property remains in the hands of chiefs; or the power of disposal lies with elders or priests, who

---

[1] Some authors distinguish between peasant farmers and the indigenous small-scale planters practising tree-and-bush cultivation in the wet forests; they call these latter *Feldbeuter* (*butineurs*, Richard-Molard, 1952), who are at a preliminary stage to that of the planter proper (*pré-planteur*, Rougerie, 1963).

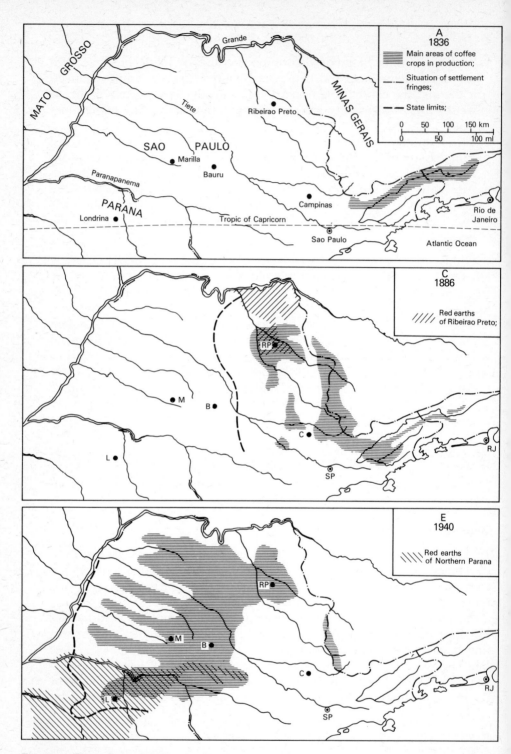

FIG. 5.10. The coffee trail in Brazil (after França 1956)

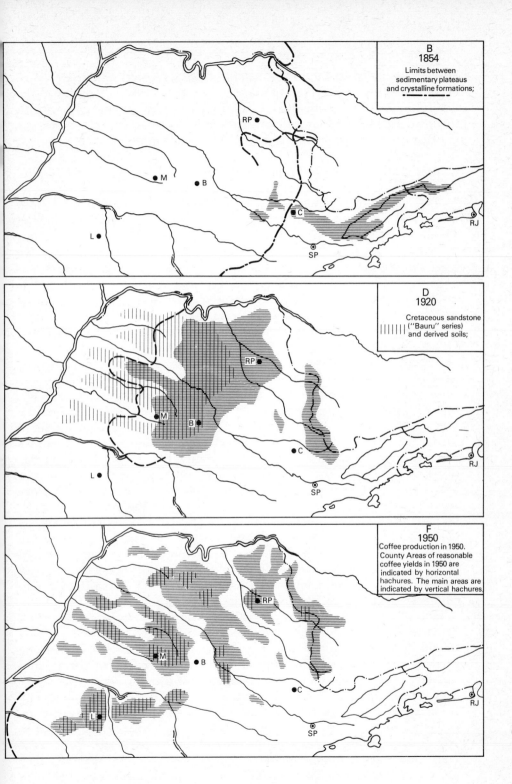

administer and distribute it on behalf of the whole community, class or extended family. Thus rights of usage generally lie with a group (often the family), not with an individual. It was only in the twentieth century, with the introduction of perennial cash crops into settled agriculture, that these old customs were observed less rigidly. They have gradually been replaced by a more individualist attitude, under the influence of production aimed at international markets. It is possible nowadays, at least in part, to acquire land by purchase.

The continuous expansion of cocoa farming in Ghana, which has been under way since the end of the nineteenth century, is a good example of the development of a tropical peasant agriculture. In the course of this development, special social and economic changes took place, which have been studied more thoroughly elsewhere (Manshard, 1961a). Much has been written about the pros and cons of peasant agriculture, with its low production costs, as compared with intensively run plantations. Many visitors with plantation experience were taken aback at the small-scale, irregular and often poorly managed peasant cocoa farming of Ghana. Other observers stress that many of the smaller and middle-sized properties of South and Central America are by no means any better run than the peasant enterprises of West Africa. The cocoa grower in Ghana applies simple cultivation methods which do not entail large capital investments, and with little effort he achieves a surplus. Even if the price drops on the international market he is less vulnerable to lower prices than the hired labour on the plantations. In times of crisis a landowner can always produce enough food for himself and his family. It is true that the irregular distribution of small cocoa farms, which are often scattered in the forest a long way from the roads, has proved an obstacle to the rational development of cocoa cultivation and the instruction of the cocoa farmers in better cultivation and processing methods. During the first decade of cocoa cultivation, large quantities of the crop rotted away every year because of transport difficulties. On the other hand, the isolation and scattered location of the cocoa farms was a great boon at the outbreak of the so-called swollen shoot epidemic, and cocoa plantations would certainly have succumbed to it faster than did the small native farms.

Besides the usual small farms, with an average size of less than 2 to 3 hectares, there are also small 'plantations' with larger cultivated areas. It should be borne in mind that most farmers, at the outset, own not one single farm but several small properties lying far apart. For private and taxation reasons only they themselves knew of their existence; they were even kept secret from friends and relatives for fear of private exploitation by the extended family. At the beginning of cocoa production in the first years of the twentieth century, there were already cocoa farms of 5 to 25 hectares. It is difficult to discover how many farms of this size still exist. Most of the larger landowners possess numerous properties in several districts; they are very reluctant to give away the exact size of their holdings, and often they have them transferred into the name of some close member of the family. The average acreage of cocoa farms in Ghana is thus bigger than is generally assumed. In southern Ghana holdings of 0·5 to 2·5 hectares are characteristic. In many cases farms of over 10 hectares can also be observed; this is specially true of the newly developed cocoa-growing areas in West Ashanti and the Volta region, where in some cases sizes reach 100 to 250 hectares. The general tendency is that the richer town dwellers—dealers, clerks and members of the professions—turn to cocoa cultivation as a capital investment, without knowing much about agriculture. The work is carried out by hired labourers, or by a tenant who gets a share of the yield. Older government officials especially, who do not think they can manage on their pensions, invest in cocoa farms in order to ensure a safe income for the last years of their lives. The future of cocoa growing in Ghana does not seem to rest either with small scattered farms or with large-scale plantations, but rather with medium-sized enterprises on which one family can cope with the work, with the additional help of some seasonal labourers. A lot would be gained if such healthy enterprises, together with government estates of larger size, could be integrated into the present system of farming.

The running costs of these small-scale peasant farms in Ghana depend on a series of factors, which in turn are influenced by natural phenomena such as soil quality and the length of the dry spell and economic influences such as the transport and market situations. Agricultural equipment for harvesting, drying and processing is simple; the main expenses are wages for the labour force; but yields are generally lower on these peasant farms than on plantations.

Under West African ecological, political and social conditions, small-holders can produce almost any crop as efficiently as large-scale production units such as plantations or state farms. Large-scale governmental production units are usually over centralised, subject to manipulation by politicians and pay wages 20 to 30 per cent higher than the open market rates. Recent studies in Nigeria have shown that government-sponsored farm settlement schemes and plantations were yielding low internal rates of return on investment while the school-leaver and smallholder tree crop schemes were yielding 10 to 20 per cent internal rates of return. This suggests that agricultural policy in the 1970s should focus on phasing out government plantations and farm settlement schemes and assisting the small and medium farmers (Eicher, 1970).

Peasant enterprises whose production is orientated towards world markets can be divided into three groups according to their origin. The first group is composed of enterprises which can be regarded as the first and almost the only producers of a particular crop and which therefore hold a monopoly of that branch of agricultural production (an example is the Indian and Bangladesh jute cultivation).

Secondly there are enterprises which originated within the framework of native cultivation, in areas formerly having an 'indigenous economy'

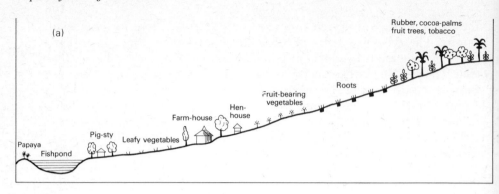

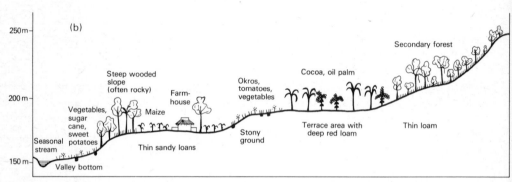

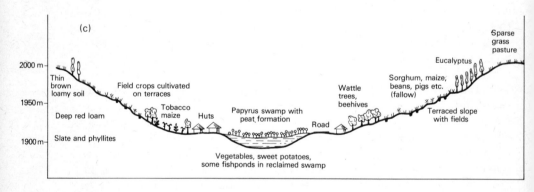

FIG. 5.11. Examples of typical sequences of land use in various parts of the tropics, presented as cross-section profiles: *catenas*

(*a*) Crop sequence on a slope in the area northwest of Singapore (after Ho, 1962)

(*b*) Diagrammatic profile of a '*huza* hide of land' in the Krobo district, southeast Ghana (after Manshard, 1961b)

(*c*) Soil and land use: *catena* in Kigezi, southwest Uganda (after Manshard, 1965b)

(cocoa farming in Ghana, as described above, is an example).

The third group comprises farming systems which came into existence when plantation land was shared out during agrarian reform movements or agrarian revolutions: the small-scale enterprises in the West Indies which

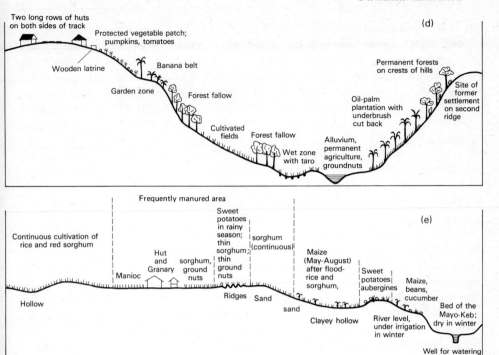

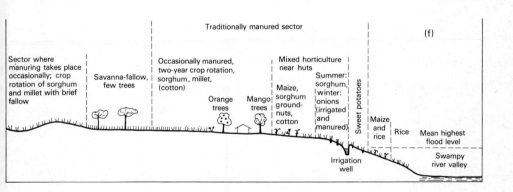

(d) Diagrammatic cross-section showing the sequence of land use in a valley in the Pene Niolo district, Congo (after Dumont, 1957)

(e) Agricultural profile of the Be district, northern Cameroons (after Dumont, 1957)

(f) Typical land use sequence in the M'Pesoba district, Mali (after Dumont, 1957)

occupy the area of former sugar cane plantations are examples. The studies by Blume (1961), on which the author largely relies for this section, provide a good insight into the markedly differentiated agricultural structures of the Caribbean Islands.

In some cases these small-scale enterprises had their origins in agrarian reforms carried out in the course of violent revolutions (e.g. Haiti 1803 and Cuba 1959/60.) In other cases they have come into existence since the abolition of slavery, a great number having been created since the beginning of the twentieth century through the state-sponsored redistribution and resettlement of uneconomic large estates. The first areas to be affected were lands with unfavourable climatic, edaphic or relief conditions, which had already been abandoned by plantation companies or which had been sold to the state for subdivision. Discordant colonial development modified types of agrarian reform, and the different administrative systems of former colonial powers make it impossible to pick out a generally valid type for all the small-scale enterprises which began in these various ways.

First, among world-market orientated peasant farms, we can set on one side as a distinct group those which came into existence in the nineteenth century. After the abolition of slavery on the former British islands (1838), part of the coloured population managed to buy up land from plantation owners. The size of the parcels varies but, on average, it was between 0·8 and 1·6 hectares. The quality of the land, however, was bad throughout, since the better lands of the abandoned plantations were acquired by the owners of big estates which were still being run remuneratively. Small fragmented private peasant farms exist therefore, alongside concentrated holdings and enlarged estates, whose areal extent is correspondingly extensive.

The kind of cultivation carried on by the small properties which came into existence through purchase or state distribution schemes depends on the region in which they are situated. In the central hill country of the islands, where here and there a subsistence economy already existed before 1838, cultivation for autosufficiency or for the nearest markets still prevails. The terrain is too broken for sugar cane cultivation and transport is too lengthy. To a certain extent, moreover, there is the danger of drought in the lower lands; although profits in the sugar business are relatively high, they could not compensate for the resulting loss of the harvest. People thus prefer to limit themselves to a varied subsistence economy in those parts of the hills where they are assured of plentiful rainfall.

On the other hand, in the coastal plains, that is in the plantation area proper, the cultivation of crops for the world market plays an important role alongside subsistence agriculture: cocoa, coconuts and sugar cane in the Windward Islands, cotton and sugar in the Leeward Islands, sugar, tobacco, coffee and bananas in the Great Antilles. Naturally, land rotation is not necessary here since the total productive area is under cultivation. In order to promote production for the world market on small peasant farms, higher profits are promised, e.g. for sugar cane production as opposed to the growing of vegetables, which would go to local markets. The marketing of export products, moreover, is well organised, whereas that of domestic

PLATE 10. Cocoa-picking in Ghana

PLATE 11. Coconut palm plantations on the coast of Ghana. This picture illustrates the close relationship between agriculture and fishing typical of many tropical coasts

products is not; this is largely due to the influence of the plantations. In extreme cases this can lead to the situation in which some islands of the Lesser Antilles rely on the import of foodstuffs from neighbouring islands (Blume, 1961).

Besides the tenants, who are more or less compelled to grow products for the world market, peasant proprietors concentrate on the cultivation of export crops of their own free will. On peasant farms of slightly larger acreage, cultivation is often limited to sugar cane, cocoa, bananas or coconuts. The small size of most holdings, however, prevents them from specialising in one crop and forces them to practise mixed farming instead, growing cassava, maize, beans and rice between the sugar cane.

This kind of enterprise—small-scale farming with polyculture—is without doubt a match for the large-scale monocultivating property, and even surpasses it; this was borne out when small farmers were settled by the state on former plantation land, for instance, in St Vincent, Puerto Rico and Jamaica. The only things that have to be guaranteed are a certain minimum size for the farm (about 2 to 3·2 hectares) and good selling and marketing possibilities. These experiences have convinced various authors (in collaboration with the former Imperial College of Tropical Agriculture, Port of Spain, Trinidad) that peasant farms could be the solution to the

PLATE 12. A cardamom field in the forests of the eastern Usumbara Mountains, Tanzania

PLATE 13. A young man cultivates his shamba of cardamom plants in the shade of some forest remnants: eastern Usumbara Mountains, Tanzania

The inhabitants of the eastern Usumbaras in the Amani area are quite well off compared to the dry lowland areas in East Africa. They are eager, however, to develop as much forest land as possible. According to the law, a cleared patch of forest belongs to the people who cleared it. This homesteader's law and the introduction of a new and very profitable crop – the spice cardamom (*Elettaria cardamonum*) is responsible for today's widely distributed small clearings.

Cardamom grows extremely well in the Usumbaras and likes the sunshading effect of some of the remaining tree giants. It needs 1·5–2 years to produce the first seeds. They are dried and sold to the cooperatives. Every three weeks about 3 kg can be harvested from a 3–5 acre plot. Cardamom sells so well that even the tea estate managers convert some estates to cardamom. People go even into the Forest Reserves. They simply ask to pay the penalty charged for this offence, which is very low. Then they clear several acres, sell the timber and plant bananas or cardamom (Private Comm., Walther, Unesco, 1970).

agricultural problems of the West Indies; their minimum sizes would naturally vary according to local conditions. It would be necessary to ban monocultures, however; what should be practised is polyculture, with an emphasis on one or two export products.

As this has been shown, the most urgent present day problem is many tropical countries lies in learning lessons from experiences with old farming systems and developing new forms which will correspond to the changed political climate. Many tropical products need better technical processing. Drying and fermenting methods are often complicated and take a remarkable amount of technical experience for granted. There is generally a tendency to a more intensive use of agricultural machinery on peasant farms in the tropics (e.g. because of new methods of spreading insecticides and artificial manures). Thus cocoa production in Ghana was raised considerably by plant protection measures and, on some selected plantations, by artificial manures (Manshard, 1962, p. 201).

To summarise: in the course of the development of 'indigenous' enterprises, competition in production and location often arises between them and plantations, with the outcome usually decided at the political level. Large-scale enterprises also differ from small peasant farms in respect of the diversification of agricultural practices; the running of the latter is still mainly motivated by self-sufficiency, and in a genuine mixed agriculture different plants are cultivated in the same fields. Yields are supplemented by raising animals (cattle, goats, sheep and pigs); this is common among almost all peasant farms in the tropics. As an example one may quote the keeping of dwarf cattle among the coconut palm groves of fishing settlements in West Africa (Togo, Dahomey, Ghana). In the same way that plantations have lost much of their world-market orientation in regions where there is plenty of demand from domestic markets, the subsistence farming so frequently alluded to has also changed a great deal. This type of economy, whose objective is autoconsumption, is now disappearing all over the world, with very few exceptions. Although on all peasant farms a great part of the necessary food supplies is produced on the spot, even in the remotest areas farmsteads are now connected with the market in some way or another. An input-output analysis of individually managed small farms (perhaps according to weight and value of production) would show that in many cases the former marked if rather crude distinction between subsistence and cash-crop farming can no longer be maintained.

The small peasant sector has considerable potential for reducing unemployment, for lightening the burden of rural poverty and stepping up agricultural production. So far only limited success has been achieved to develop this economically underprivileged group which together with landless labourers and unemployed make up the mass of the rural population.

The size of this development programme is immense. About one-third of the total population of the developing world (about 700m to 800m people, excluding mainland China) belong to this economically deprived part of mankind. Among other agencies, the World Bank has tackled this problem. Since the early 1960s the most notable trend in the Bank's lending for agriculture has been the diversification beyond basic irrigation infrastructure into on-farm activities, technical services and related rural development. Increasing emphasis is being given to rainfed agriculture, to storage, marketing, seed multiplication, forestry and fisheries projects. It is also hoped that smallholder development schemes, including the settlement of landless people, can be extended to support larger numbers of small farms, providing them with additional income at the lowest possible cost per farm family.

The number of projects aimed at producing food crops, livestock and fish for local consumption should increase as individual developing countries become more conscious of the need to provide employment opportunities and distribute incomes more equitably. Greater emphasis is likely to be given also to projects intended to benefit small farms and to be labour-intensive (cf. *World Bank Sector Study*, 1972).

Taungya forestry systems

Combined tillage and forestry is a specialised form of agricultural land use in the tropics. In many ways this is a kind of permanent cultivation, since crops are grown for several years between the seedlings of commercial export timbers such as teak, mahogany and limba. This combined cultivation, also known as *taungya*, was developed by Brandis in Burma (see p. 61). The aim of the technique is to curb shifting cultivation, which is harmful to productive forestry and inadequate in output, and also to regulate it. At the same time the basic diet of the population is improved, a factor of special importance in densely populated areas. A parallel improvement in social and health conditions is therefore also obtained.

To have any promise of success, combined cultivation is only possible near areas of concentrated population where land is scarce; in such areas the prospects of good harvests provide a stimulus, and the labour force may also undertake additional duties, such as clearing the ground around the tree seedlings. However, areas chosen by forestry commissions for combined cultivation are usually secondary forests which no longer contain commercial timber; in some cases, as in Tanzania, this is the way in which reafforestation is carried out. Combined cultivation can be further described by means of a few examples.

In Nigeria where this method was practised in the vicinity of Sapele (Benin Division) the forests had lost their valuable timbers because of slash-and-burn agriculture. At the same time rubber and oil palm plantations had reduced the amount of land which could be used for individual cultivation. Every year an exactly defined part of the forest reserves was released for *taungya* and distributed to individual communities in so-called *taungya* series. Each family obtained about 15 hectares free, to cultivate for their own needs. In return, the farmer had to prepare sections of the forest for planting, put the seedlings in the ground, weed them and prune them. After a few years the forestry commission took over the section and whatever work remained to be done was carried out by forestry workers, while the farmer was allocated a new piece of land which he again cleared himself and prepared for cultivation. The principal crops were yams and manioc, and to a lesser extent bananas, melons and vegetables. The main commercial timbers used were various kinds of mahogany (*Gmelina arborea, Cedrela mexicana, Khaya ivorensis*), teak and *Terminalia*. In the Benin Division about 4 000 hectares of forest cultivation had been established in this way by the end of 1960 (Hesmer, 1966).

In Zaire mixed cultivation of limba and bananas is important, mainly taking place in the Mayumbe district on the lower Congo. Combined cultivation was introduced here by the company called Société Agrifor de Lemba; having exploited the existing stocks of limba by the end of the 1930s, the company developed the *sylvobananier* system. Here the farmers had the land for a few years for the cultivation of export bananas. After one or two years the limba saplings were planted. Banana harvests were remunerative for about five years, then the section became pure limba forest. Later, promising attempts were made to grow cocoa or coffee under the shade of the limba trees, after banana harvesting had finally terminated. Thus it has been possible, through the simultaneous cultivation of bananas and ground crops of cocoa beneath Limba trees, successfully to expand combined cultivation into a continuous double cropping of timber and plantation products.

In Tanzania combined cultivation was already carried out for a while during the German colonial period. Pine, cypress and eucalyptus are the main commerical timbers used, since they stand the climate of the highlands best. The same applies to Kenya. Particularly interesting is the cultivation of pyrethrum between pines on the Meru. Contour ploughing is necessary on sloping terrain to prevent soil destruction.

Hesmer (1966) has shown in detail the importance of combined cultivation and forestry for both the forestry industry and the economy of African and Asian countries. The advantages of *taungya* are manifold, ranging from higher yields in agriculture and forestry to protection from soil erosion in savanna regions (Togo, Ivory Coast) and more regular employment and earnings for the population. But discipline among the farmers, expert knowledge on the part of the forestry people, and good organisation in the forestry offices are decisive for success.[1]

Pastoralism and grassland utilisation

A survey of the various types of tropical pastoral economy reveals a striking contrast between the traditional nomadism of the dry belt of the Old World and the stock farming of the New World and the southern hemisphere. In the case of nomadism in the deserts, steppes and savannas of

---

[1] Although in most countries the *taungya* forestry system still seems to fit quite well the stage of shifting cultivation or land rotation, in some parts of India further development has come to a deadlock because of lack of enough land. In this case the system has gone beyond the limits where the *Caboclo* stage (Valverde, 1971) still prevails.

the Old World, individual land ownership and tillage is not usually known[1]. There is a great demand for space, varying according to climatic conditions and the basic fodder requirements of cattle, sheep, goats or camels.

Richthofen's definition (1908) of the term 'nomadism' contains the characteristic attribute of the whole tribe migrating with the herds, but he does not mention agriculture. However, Herzog (1963) points out that this 'pure' type of pastoral economy is rarely found; on the contrary, there are hardly any nomads who do not have some relationship with agriculture. On the whole the distinctions made between various types of economy, for the purpose of characterising separate groups within the theory of economic stages, are usually far too simplified. It is true that there are families and groups who belong to only one type of economy; but among a whole people a combination of different forms is usually found. It is also interesting to note that these types usually only characterise the economic structure, rarely the social structure.

Free-ranging nomadism, even though limited by few boundaries, is now often circumscribed by wire fencing and by a subdivision of the area into separate pastures, which are all used in turn. This system allows some control of pasturing and choice in breeding, and ensures better soil protection.

In the New World nomadism is absent. Already during the period of colonisation of Latin America large landed estates had been developed with an emphasis on animal husbandry (stock ranches, *estancias*, *haciendas*, *fazendas*, *hatos*) and mainly producing meat and hides. The pastures were often over 1 000 hectares in area. Dairy farming only gained importance later on, in the vicinity of towns. The *hacendado* and his *vaqueros*, the *gaucho* and the cowboy represented ways of life which in their economic thinking were different from that of the arable farmer.

In order to be able to describe the spatially different types of pastoral economies, mention has to be made of the system of rotating pastures, which may or may not be related to climate. Apart from the stock-raising areas of the New World, South Africa and Australia, where the easily moved electric fence is common, rational methods of rotating pastures or moving stock from one region to another in a progression from breeding area to large-scale slaughterhouse, have made little headway in the tropics so far. Simple ways of rotating grassland by tethering (fastening each animal to a peg) can, however, often be observed.

From the terminological point of view, the following distinct types of pastoral economy exist in the tropics: nomadism proper, partial nomadism, transhumance and livestock raising on mountain pastures. There is an essential difference between nomadism proper and the other types: the

---

[1] In an essay on the human geography of deserts Dresch (1966) referred to the important economic structural changes taking place in the desert belt of the Old World, such as the decline of nomadism and the development of dry-farming and irrigation.

latter have permanent settlements, inhabited all the year round, and a modest agriculture or the cultivation of perennial plants. These forms occur with both plough and hoe tillage. Transhumance is a partly nomadic animal husbandry combined with the driving or transport of stock between two (or sometimes several) pastures, which can only be used seasonally and which are quite distinct as regards altitude, climate and vegetation (Hofmeister, 1961). The permanently inhabited residence almost invariably lies near one of the two pasture areas and is not necessarily organically connected with the use of the pasture. In mountain livestock raising, animals are kept in stables over the winter and fed on hay; this does not apply to transhumance, as a rule. In lower latitudes the various types of transhumance are mainly the product of alternating rainy and dry seasons. This tropical transhumance, which follows the precipitation cycle of the year, moves downhill in the rainy season and uphill during the dry spell. Whereas in the Mediterranean region transhumance is specially restricted to sheep, in the tropics cattle migration prevails. Descriptions of such seasonal cattle migrations have been given for tropical South America —for instance, in Colombia, Ecuador, Brazil and Venezuela (Wilhelmy, 1954, and for Central Africa (Rhodesia).

In contrast, a form of transhumance in which stock move higher during the rainy season predominates in, for example, parts of Amazonia, where the cattle herds have to avoid the flooded lowlands. Occasionally one finds simultaneous movements in both directions (for instance in Ethiopia) cattle are driven lower and sheep and goats higher when there is sufficient fodder (Wilhelmy, 1966). Cases of transhumance in several stages have also been observed, e.g. among the Galla in southern Ethiopia (Haberland, 1963).

In order to illustrate the different types of pastoralism one may quote a few examples. In tropical Africa the Fulbe (Fulani) and the Masai are particularly well-known cases of typical nomadic peoples; the latter have preserved their pure nomadic herding tradition almost intact up to recent years.

A century ago, at the time of their greatest power, the warriors of the Masai (*morans*) dominated the steppes of the East African Rift Valley and the neighbouring high plateau. During the dry season they drove their herds up to the highlands, which received more rain and where the grass was richer. In the rainy season the cattle grazed on the grasslands in the bottom of the East African Rift Valley. Under British rule the Masai lost a large part of their grazing lands and were reduced to a small reserve. This reduction of their living space, combined with the increase in population and livestock, resulted in the grasslands being overgrazed. In 1961 the Kenyan Masai, with 950 000 cattle and over 650 000 sheep and goats, surpassed the capacity of their grazing lands by 100 per cent. The traditional basic diet of the Masai, which consisted of milk, blood and meat, is supplemented nowadays by millet, wheat and other vegetable foods,

FIG. 5.12. Migrations of camel nomads in the Sudan; Central Kordofan (after Born, 1965)

The camel nomads of central Kordofan (Republic of the Sudan) can be subdivided into two groups. The northern group is mainly composed of the Kawahla, Kababish, Hawawir and Beni Gerar, with their homeland in the north of the Qoz Zone.

The rainy season occurs in the months of May to September. In mid-May the northern tribes migrate into the central Qoz zone, where the first rains start; in June they leave again and wander northwards following the movement of the rainfall. The return of the nomads from the north begins in September; in October they once more reach their homeland north of the Qoz zone. The 'Gizzu migrations' only take place in very wet winters, when the dune areas on the lower course of the Wadi Howar have an adequate grass cover.

The southern tribes too follow the rains northwards, although this hardly ever takes them across the northern boundary of the Qoz zone. Only the Shenabla move south, to the refuge of the Nuba mountains, before the dry season begins, since they do not have any wells of their own on the northern fringe of the Qoz zone. They stay on the pastures of the Nuba mountains from January until the beginning of May; then, with the onset of the rains, there follows the return migration to the north.

These migrations of the camel nomads always take place in small groups and not as a whole tribe, since in the latter case there would be a danger of over-grazing.

The Qoz zone is an extensively-used area of peasant settlement with light sandy soils. The inhabitants are, in part, nomads of various tribes which have become sedentary.

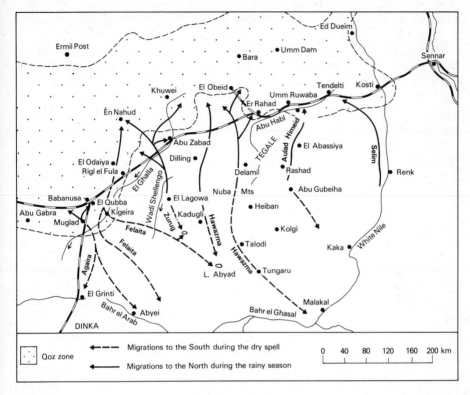

FIG. 5.13. Migration of cattle nomads in the Sudan: Nubaland (after Born, 1965)

The areas of origin of the cattle nomads lie south of the Qoz zone. They are Arab tribes which, on penetrating into these regions about 200 years ago, changed over from camel raising to cattle breeding. The original negroid population was driven into the mainly inaccessible mountain areas. Thus the homeland and the pastures of the Baggara, the cattle nomads, extends southwards roughly from the southern fringe of the Qoz zone to the Bahr el Arab and the Bahr el Ghasal.

The southwards migration, from December to May, takes place for similar reasons as with the camel nomads – to remain near the water sources of the homeland during the dry season would result in a severe degradation of the grazing. The nomads accordingly take refuge in the vicinity of Lake Abyad and further south; sometimes they even cross the Bahr el Arab. Migration to the north, on the other hand, takes place rather to 'avoid' the rainy season, because the heavy rainfall from June to October transforms the clayey soils south of the Qoz zone into hardly traversable mud. This, and the insect plague which starts with the beginning of the rains, forces the nomads to migrate north into the Qoz zone.

The cattle nomads, too, prefer migration in individual groups rather than as a whole tribe. Some of the cattle nomads became sedentary and changed over to agriculture, nowadays mainly cultivating cotton.

which they buy or barter from neighbouring tribes. In all the core areas of nomadism in the Old World a change can be observed, accelerated by political educational measures; this is leading away from the former subsistence economy; Born (1965) offers some illuminating examples of the migration of camel and cattle herds from the eastern Sudan (cf. Figs. 5.12, 5.13)

Rotenhan (1966) draws a vivid picture of the problems of cattle-raising in Sukumaland (south and southeast of Lake Victoria) in Tanzania. The agriculture of the Wasukuma serves two purposes: it ensures their vegetable food needs and it procures cash income. Animal husbandry, on the contrary, which is carried out on a large scale, is not so much a source of income as an insurance for the family and a means of gaining prestige in society. Cattle are bartered and their products are consumed. The milk is drunk, processed into cheese or butterfat, or occasionally sold or exchanged; but generally the words used by the Wanyaturu (central Tanzania) could apply also to the Wasukuma:— 'Cattle are our banks, our granaries, our fields, our women, our families. They are everything to us.'

Now, as always, the esteem in which a family is held rests largely on the number of cattle they possess. Although commercialisation has made big inroads and money is known as a means of exchange in even the remotest regions, they still cling to traditional concepts of values and keep as large herds as possible. The peasants have not yet realised that social differentiation can be far more effectively based on the size and quality of cultivation than on the possession of cattle. Thus a considerable portion of the profits from cotton cultivation is used to acquire additional cattle and not for profitable investment or the purchase of goods and materials for production.

Moreover, the Sukuma regards the increase in his stock of animals as the best insurance against economic and social risk. He has to protect himself against the uncertainty of rainfall and the destruction of his harvest by pests and plagues. To store larger reserves of food seems pointless in view of inadequate storage facilities and high losses through pests. Instead, any surplus from agriculture is converted into cattle by sale or exchange. The Sukuma resorts to this accumulated 'capital' when the family needs money, when the harvest fails or when money is needed for the education of the children. Apart from that, the Sukuma herds cattle in order to enjoy a secure old age.

The Wasukuma, as well as most of the other tribes in East Africa, base a system of human relationships and security on the bestowal of cattle. Among other practices the 'bride price' (*kukwa*) is well known: the bridegroom or his family give a number of cattle to the bride's father at marriage. Should the behaviour of his daughter not conform with the acknowledged code of custom, the bride's father has to return the animals. On the other hand, the bride's father can fetch his daughter back, and keep the cattle handed over to him, if the bridegroom maltreats his wife. Thus the transfer of animals is a means of keeping both parties interested in the success of the marriage. On the payment of the bride price the husband's family simultaneously acquire rights over the offspring of the new marriage. Rights over children are important from the point of view of future labour and care in old age. In Sukumaland there are no fixed rules as to the size of the bride price. The more extensive the cattle stock of an area and the more respected the bride's family, the more animals are handed over.

These social functions of stock-raising cause the Sukuma to hoard cattle and forgo part of the potential meat yield. If need arises, only poorer quality animals are sold. Occasions for slaughtering cattle are usually big family events such as births, weddings or deaths. A neighbour's help in building a house or cultivating a field is usually rewarded with a 'slaughter feast'. Apart from these occasions beef is only consumed when cattle have to be slaughtered or have perished. The consumption of beef, therefore, is not so much the result of a calculated use of livestock as the consequence of an unfortunate loss.

Whereas profound changes have taken place in the agriculture of the Wasukuma, the principles of stock-raising organisation have hardly been altered at all. It is true that stocks have been increased, but without exploiting the potential productivity which this extensive animal breeding offers. As formerly, the social functions of this branch of economic activity still have priority.

There is no local or tribal institution responsible for the organisation of the communal use of grassland, or which concerns itself about maintaining the productivity of the pastures, the control of erosion or resowing. Everyone has the right to allow his total stock of animals to graze on the communal pastures. Of necessity, out of this situation there results an inconsiderate increase in the number of stock, because the bigger one's own herd, the more advantage is gained from the communal land. Thus, the pastures of the Wasukuma are overgrazed and only produce a small fraction of what could be obtained if a proper organisation existed. Moreover, areas in danger of erosion are still recklessly used until the grass cover is completely destroyed.

The shortcomings of cattle-raising in Sukumaland are in no way the result of natural conditions being unpropitious, as often is the case with pure or partial nomads. They are caused rather by a failure to realise the possibilities which exist, by social ties, and by the inability of the individual farmer to carry out essential changes of his own accord. Essentially, only one isolated facet of market orientated farming has yet been grasped: the possibility of earning money from cotton cultivation.

Naturally, one should not simply accept these stereotyped and possibly too rigid ideas. In northern Ghana, where similar conditions prevailed until recently, many African cattle herders have learned to adopt a strictly economic attitude, in the modern sense of the word (Hill, 1970).

An interesting variant is 'commissioned' cattle-raising, as practised in southern West Africa by Fulbe herders brought in from outside (Manshard, 1961a, 1965b, Müller, 1967); it is common, for instance, in southern Ghana and southern Togo. The Fulbe are employed to look after large herds. As a reward they are given the right to sell milk, butter and cheese from the herd.

In contrast to Africa, the Near East and Central Asia, a pastoral economy has never been developed in southern Asia. In India, which has the largest cattle population on earth, Hinduism restricts their use, in large part, because of the prohibition on slaughtering them. Of India's estimated 250 million cattle about half are sickly. They consume food but they yield no milk, they are unable to draw the plough, and are thus a burden to the economy. In Southeast Asia, on the other hand, the water-buffalo is a tried and versatile working animal which has become indispensable in the ricefields.

The question of how nomadic groups may be converted to sedentary life is especially open to discussion in the high mountain regions of Asia, such as Afghanistan. Too hurried moves towards the abolition of nomadism seem to be dangerous. On the other hand, there is a need for nomadism to be guided towards a kind of economy more in tune with modern conditions. Some proposals for the 'modernisation' of such nomadic groups which could be realised at relatively low cost, may be summarised as follows:

1. provision of a veterinary control service, whose field of activity should also include the instruction of nomads in the place of hygiene in animal raising;
2. health control over people as well as animals by the installation of 'sanitary sluices' on migration routes, such as sometimes exist already on frontier crossings;
3. regulation of the grazing rights of individual nomadic groups and the introduction of periodic closed seasons to ensure the regeneration of pastures;
4. the promotion, by creating cooperative organisations, of the marketing of animal products and handmade articles such as earthenware and woven cloth;
5. the setting up of livestock insurance companies as a protection from disasters (these have been successfully created on behalf of nomadic groups in North Africa);
6. raising the educational standard of nomads by increasing the number of mobile schools.

In the arid regions of the marginal tropics (Southwest Africa and Northern Australia) large-scale farms with extensive acreages (in some cases 10 000 to over 100 000 hectares) and large herds of up to 10 000 head are common. The number of animals and labourers per hectare is low; on the other hand, the costs of fences, waterholes and so on are high. Production has only one objective (meat or wool) and diversity is not a consideration in these marginal locations. Only rarely is the cultivation of fodder worth while.

In North Australia, some of the big cattle stations cover areas as big as Luxembourg or Belgium. The cattle densities (often under two head of cattle per square kilometre) are lower than in the subtropical zone. However, these figures do not mean much because under the open-range system many remote areas are not grazed in the dry season. Open-range implies that there are no fences, often not even between neighbouring stations. For the management of pastures and the breeding of cattle, this has serious drawbacks. In fact, under a semihumid to semi-arid climate, conditions are rather similar to other pastoral regions of the tropics; i.e. in the rainy season cattle herds are spread over wide areas, while in the dry season they concentrate round water holes. Also in north Australia (actually just inside the State of Western Australia), mention should be made of the great Ord River irrigation scheme, at present amounting to some 12 000 hectares of cultivation, upwards of one-third of this area being under sorghum for cattle feed.

The scientific progress made in this technical age has clearly been of most help to sedentary peasants in waging their age-old political conflict with the nomads, to whom they used to pay tribute. Grazing areas are more and more reduced in extent by the peasant cultivator. With increasing population pressure, problems of overgrazing result, and with them the destruction of soil and vegetation. So far there are even fewer signs of progress in tropical pastoralism than in shifting cultivation. Systematic breeding and the fight against pests are difficult to carry out and very costly. An improvement in fodder supply is often impossible in the savannas, since the nutritional basis is insufficient, unless a changeover to sedentary cattle raising takes place. Such a move towards a sedentary system, and its transitional stages of half and partial nomadism, calls for the construction of wells, irrigation systems and settlements.

Outside the agricultural areas which have been under European influence, an integration of tillage and animal husbandry is lacking. Intensification would be particularly worth while in more industrialised areas and in the vicinity of towns. In this connection reference should be made to the research carried out by Otremba and Kessler (1965, pp.134–47); they have investigated the place of animal rearing in the agricultural regions of the world and within various economic, natural and cultural systems.

In the spatial system, which is subdivided into town and countryside, market and production point, the economic problems of animal husbandry begin on the fringes of large cities and on the periphery of industrial concentrations. This kind of development, near large centres of consumption, has not progressed very far yet in the tropics. It is only thanks to modern transport and conserving techniques that less favoured areas of stock-raising (e.g. the wet forests of the tropics) have been connected to more intensive ones; and even the latter are often far from markets.

Animal husbandry in the tropics is concentrated in the savannas and xerophytic forests, where natural pastures exist. In the more humid regions of the tropics, livestock raising is increasingly fixed in location, often in close relationship with the cultivation of millet, maize, sugar cane or coconut palms, whose waste products serve as fodder.

In connection with the improvement of animal husbandry, attempts are being made in many tropical countries to establish dairy industries in order to provide more animal protein and thus improve existing malnutrition. There are only a few tropical regions where the provision of fodder for an intensive dairy industry does not depend upon the availability of water throughout the year. For most tropical and subtropical countries this is an important problem. Dairy animals of low productivity have to obtain most of their fodder by grazing, thereby often damaging the existing vegetation. Irrigated cultivation of grasses and fodder legumes is expensive and often competitive with other food or cash crops (Whyte, 1967).

Finally, in large economic-cultural regions, historico-political and religious-spiritual forces influence the form of pastoral economy, finding their obvious expression in the various ways in which cattle-raising peoples, from nomads to ranchers, live and run their economies.

The sites of pastoral settlements are mainly influenced by water supply conditions. In particular, readily obtainable ground water, as well as episodically or periodically flowing streams, attract temporary settlements of migratory herders. The location of the seasonal settlements of half-nomadic cattle breeders is also strongly tied to climatic (winter quarters) and relief conditions. Formerly, fortification or defence played an important part. Occasionally a transition to cultivation or an oasis economy during the rainy season can be observed. One prerequisite for a changeover from a nomadic to a sedentary life is the introduction of a systematic pasture rotation, which will incorporate seasonal settlements. Many groups of nomadic herders have a highly differentiated social class system, which is reflected in the shape and arrangement of their tents and huts. Special facilities are necessary to provide night shelter for the animals, since they have to be protected from beasts of prey. In the case of the southwest Bantu (e.g. Zulu, Swazi) the typical beehive huts are arranged according to the age of the inhabitants around the animal kraal. Among the East African herding tribes, oblong dung-covered domeshaped huts are erected, again grouped around the animal kraal. The camel-raising Turkana of the far north of Kenya live in kraal settlements which resemble those of the Masai and other nomadic groups. Their huts, however, are not covered by cow dung; instead, they put skins over them during the short rainy season. Almost every one of these family settlements shelters 25 to 30 cattle, about 10 camels and around 100 sheep and goats.

In areas influenced by European colonisation, where a widespread pastoral economy has developed, cattle ranches, sheep farms or *estancias* are typical. The buildings of the stockfarmer in the South African Karroo usually consist of a one-storey house, furnished like a town house, belonging to the owner or manager; small houses for the labourers; sheds for sheep-shearing and storing wool; garages; and stables for small stock (pigs and poultry). Wind-motors, providing water and electricity, also belong to these settlements. Such a settlement is usually surrounded by groves of trees (eucalyptus or conifers) and small gardens. Similar settlements can also be found in other stock-farming, pastoral regions with analogous structures (e.g. in Southwest Africa).

Game cropping

Another variant of grassland utilisation is 'game cropping'. Proposals, to exploit systematically the game stocks of the tsetse infested, thinly populated plateaus of East and Central Africa, have been studied for some years. The aims of this 'wild life farming' or 'game cropping' may be summarised briefly as follows:

1. Care of, and in some cases breeding of wild life species (e.g. antelope in Rhodesia); selective hunting and slaughter for domestic meat consumption; as a result, better protein diet for the population; export of skins and furs.
2. Making use of areas of poor vegetation as habitats for wild life farming could in many respects be more advantageous than cattle-farming. Soils would be protected and the dangers of overstocking avoided. Whether these proposals can actually be realised and worked out economically has yet to be studied thoroughly. Difficulties arise in hunting the animals, in transport to the slaughterhouse, in deep-freezing the meat and in solving the obscure land tenure conditions of many wild life reserves.
3. Increased care of wild life stocks would stimulate tourism, which is extremely important for developing countries in its provision of foreign currency.

Erz (1967) provides an excellent summary of the problems involved. Whereas economists have only rarely, or not at all, come to terms with the economic aspects of wild life, experienced conservationists somewhat enthusiastically extol its advantages from the point of view, not only of ecology, but also of management. As propounded by ecologists, the strongest argument for integrating game cropping into the agrarian structure of stock-raising regions is its importance for the ecology of the countryside and the conservation of nature. Starting from the biological characteristics of the individual animal, the ecological and physiological performances of wild and domesticated animals are compared. The significance of every single feature can be compared from the point of view of economic policy and business administration. The essential facts, as they have so far been stated, may be recapitulated briefly as follows.

As compared with the monoculture of cattle, a population of wild life species uses a far greater variety of plants for its food than do cattle and other domesticated animals. Besides different kinds of grasses and other ground cover, roots and the foliage of bushes and trees are consumed. Feeding on woody plants is of special importance on savanna pastures since the greater part of primary vegetable products there consists of such material. If only cattle-breeding were carried out, only a fraction of these vegetable products (that is to say, the grass cover) would be transformed into meat. The 'vertical' utilisation of pastures is a characteristic of wild life communities, so that a greater variety of species and a greater density per hectare are possible, compared with cattle. Studies have also shown that a greater biosubstance is achieved (i.e. a higher production of meat, expressed in kilos per ha) than with cattle or other domesticated animals.

Such a superiority in meat production in marginal savanna conditions would be important economically. A further economic advantage is that a wild life population spares productive regions far more than do grazing cattle. Since feeding habits embrace a much larger range of the foodstuffs available on natural savanna pastures, as wild herds cover a greater area in their wanderings, and as they are much freer from dependence on watering-places, pressure on pastures and their ecological structure of edaphic and hydrological factors is far lower than in the case of cattle-raising.

Using such arguments, at least the bases of discussion have been created for the eventual integration of wild animals and wild life farming into the agricultural land use of the African savannas.

The value of production of wild animals may also exceed that of domesticated cattle. The parts of wild animals which can be used (i.e. the proportion of dead weight to live weight) is between 50 and 67 per cent, whereas the values for cattle usually range from just under to slightly above 50 per cent. According to studies in East Africa, wild animals compare very favourably with cattle in their yield of first-class, pure lean meat: wild animals yield 41–47 per cent, cattle about 33 per cent. In this estimate of relative performance, the relatively higher utilisation of proteins in wild animals slaughtered also plays a part. The decisive feature of the insufficient diet of the African population is a lack of animal protein rather than a simple lack of food. Moreover, these higher performances in wild animals can be achieved at a much younger age than in cattle, since increase in weight is generally faster in the growth of wild animals.

The argument among scientists is that these points of view ought to convince the individual farmer that, in view of its poor prospects of profit and the damage it does to productive areas, they should at least cut down on cattle-raising in favour of game cropping, even if they do not give it up altogether. Further stimuli to the maintenance of wild life stocks for man's direct advantage should be the lower capital investment needed for wild life farming (breeding stock not required; no need to open up new land or improve it by fences, water, etc.; no expenditure on personnel).

Farmers are nevertheless hesitant to adopt game-ranching; their argument is that relying completely on wild life is an unsafe way of providing an economic meat supply and, from the point of view of managing an enterprise, an uncertain source of income. They are especially hesitant about a complete changeover to a system which has not yet been fully tried out anywhere. In contrast to 'game ranching', which is the name given to the utilisation of wild herds on relatively extensive acreages, 'game farming' may constitute a safe, if subsidiary source of income from wild life, as found, for instance, on various farms in the Transvaal, where it often started as a pastime.

The biological principles of sensible game cropping, without damaging game stocks, are the following: only surplus stock should ever be 'creamed off', that is, stock which would perish anyway through being decimated by natural factors (enemies, diseases, food shortage, and so on). In game cropping, the quota of animals to be shot should lie within a safety margin. Game cropping will mainly be carried out on soils with marginal yields. The occasional permits granted to practise game ranching within reserves should be exploited as economically as possible, a point on which the game wardens should in any event insist. The extent to which price and market manipulations come into the business, however, lies beyond the sphere of influence of conservation.

In a schematic gradient from wilderness to cultivated region, game-cropping can be inserted at varying intensities. Thünen's intensity zones in economics can be applied to game farming, too. From the point of view of conservationists, a form of conservation which is based on man's economic instincts would be the safest guarantee for the permanent conservation of African wild life stocks (Erz, 1967).

Problems of agricultural mechanisation with special reference to tropical Asia

Agricultural implements, whether used by hand or drawn by an ox team, have been employed since early times by individual tropical peoples in the working of their land, including, for instance, the oldest agricultural tool, the digging-stick, from which the spade developed, and the forerunners of the plough. Agricultural mechanisation proper—the utilisation of motor energy for agricultural purposes—only started more intensively in tropical countries after the Second World War. Tractors left behind by armies often constituted the initial technical equipment, and stimulated further mechanisation. Plantations or large-scale enterprises in European hands were the first partly or full mechanised properties in the tropics. In Malaysia the large foreign plantation companies, which began to import tractors as early as 1930–31, were the pioneers of agricultural mechanisation.[1]

[1] Questions linked with agricultural mechanisation have been mentioned several times in previous chapters: for example, irrigation devices, and the provision of rice mills, drying sheds and oil presses, which can be described as 'mechanisation' in the broad sense of the word. In this chapter there will be special emphasis on the importance of the tractor for working the land and for agricultural development in general (see also Gallwitz, 1963, Kisselmann, 1965, and Bergman, 1966).

Managerial, technical, financial and also social problems put a brake on the drive towards mechanising. Doubts have cropped up from time to time as to whether it is desirable, economically and socially, to develop agricultural techniques by leaps and bounds. Different measures have been undertaken in individual regions according to the different agricultural structures. In India, for instance, the Planning Commission envisaged 10 000 tractors per year being built in the country, from 1965 onwards. Simultaneously, a department of the Central Ministry of Agriculture was working on designs for new improved wooden ploughs. It is thus characteristic of a large part of the tropics to find modern machinery side by side with the oldest manual procedures, harvesting and transporting sugar cane by hand, for example, and extracting the juice by means of a stationary tractor. This co-existence is also reflected in the way different agrarian social stages of working the land may be found side by side (e.g. in India, share tenancies alongside properties mechanically cultivated under the landowner's management). In Pakistan also mechanisation is in progress. There are about 40–50 000 tractors, of which most were imported fairly recently; most of them are used on farms of over 10 hectares (Mullik, 1972).

Within a single country, although separated spatially, stages of mechanisation can vary. Thus, for instance, the large rice plantations of Sumatra are often fully mechanised, whereas the mechanisation of rice cultivation in Java is out of the question, or is at least attended with considerable difficulties owing to the small size of parcels, terracing and above all the lack of capital among the small-scale farmers. Almost everywhere in the tropics mechanisation is still in the early stage of the heavy work on the land being changed from animal traction to motor power, exception being made of plantations, which differ greatly from the rest of agricultural enterprises in their industrial organisation, forms of mechanisation, and their social connotations. The tractor is predominantly used for clearing waste land, terracing, ploughing and cultivation. Other activities such as sowing, irrigation, the tending of crops, manuring and harvesting, are still largely carried out by hand or by a team of oxen. Mechanisation is thus only partial and not complete.

There are numerous different kinds of mechanisation and they are usually found side by side; their importance varies, however, in individual countries. Tractors and trailer equipment are usually made available in large numbers by state machinery stations, state and privately owned large-scale properties, and by cooperative organisations either working the land in common or lending the machines to their members. To date, the greatest proportion of the equipment is in the hands of private and state large-scale holdings, and only a small percentage belongs to agricultural societies. The great mass of small-scale peasants have to get along without tractors.

The tasks performed by agricultural machinery are similar in the tropics to those in other agricultural regions; but in the tropics, especially in the densely populated rural areas of tropical Asia with their hidden unemployment, the need to economise on human labour does not seem urgent, although it is a factor which has considerably accelerated agricultural mechanisation in the industrial nations of today. Again, in the case of the latter, the purpose of introducing the tractor, together with the appropriate accessories, is to raise labour efficiency as much as possible, to create sounder bases for high yields and to help to improve the quality of harvests. But both the prerequisites for, and the effects of, mechanisation vary greatly in different agricultural regions.

There have been setbacks to the process of mechanising agriculture, caused for instance by the wrong use of tractors in tropical cultivation. Topsoil and soil fertility are far more easily endangered in the tropics than in temperate climates; at high temperatures, organic matter decomposes faster and the soil loses some of its properties; violent rainfall washes away much of what remains. Wrong use of tractors can increase these dangers. The essential requirement in employing tractors is that at all times the ground should be reasonably dry. If attention is not paid to this factor, severe structural damage can be caused to the soil. In the temperate zone frost is a big help in avoiding damage of this kind. In the tropics, however, with frost absent, the disabilities caused by a deterioration in soil structure can last for years. Furthermore, agricultural implements which could at least prepare the soil for cultivation, substituting the action of frost in putting it in a ready state, do not usually exist.

It is difficult to calculate the extent to which mechanisation will increase harvests and rentability; this depends largely on local conditions, such as the size of holdings and fields, and the crops cultivated.

There is a considerable possibility of increasing the intensity of cultivation through only partial mechanisation, as has been done, for instance, with upland rice in Indonesia and Malaysia, among other countries. Apart from very few exceptions, fields are only tilled once a year. Cultivation is not begun until the first rains have soaked the soil enough for it to be worked with draught buffaloes or oxen. If it is possible, thanks to the use of stronger mechanical motive power, to prepare the soil and seed it before the beginning of the rains, the rice shoots can already start sprouting immediately after the first rainfall. These young shoots are thus well ahead in their growth (by about two months) compared with others cultivated in the old way, and it is possible to obtain a second rice crop or grow another follow-on crop such as maize or soya beans. In the mechanical preparation of soils for wet rice cultivation, however, care has to be taken that the soil is not worked too deeply; otherwise, loss of water will lower the yield.

It is not uncommon that the use of tractors also influences the quality of the crop. On rubber plantations, for instance, great importance is attached to the rapid transport of the latex; and in cultivation generally there is a better chance of bringing in the harvest at the right time.

Some considerations about mechanising the work, such as quicker cultivation of the fields, better time control of processes, greater control over quality of labour, seem to make partial mechanisation desirable in the tropics, in certain circumstances. The profit attached to the possession of a tractor is increased if it can also be employed for special jobs such as irrigation. The main goal of agricultural mechanisation is to intensify cultivation on land already under crops, and also to reclaim new land; this can be achieved through partial mechanisation. In Malaysia, for instance, until recently 8 000 to 12 000 hectares of rainforest have been cleared annually by hand. More machines will be used in future and clearing will be made considerably easier. Large and hitherto untapped productive reserves can be opened up for agriculture and new settlements (Kisselmann, 1965).

Especially in this context, the effect of mechanisation on labour demand can be positively assessed. It has to be taken into account that the number of work-people basically depends more on the intensity of cultivation than on the new use of tractors. Where agriculture was carried out intensively before, the numbers employed may remain stable or drop only slightly. In cases of extensive agriculture, however, or where land has been left uncultivated, the employment of machinery may perhaps mean that more people can be fed and rationally employed. This is particularly true when wasteland is brought under cultivation. In tropical countries a great proportion of the rural population, landowners as well as landless labourers, are underemployed, working only for a few weeks in the year. For example, in some villages in the federal state of Madras (southern India), the inhabitants were found to be unemployed for between 53 per cent and 81 per cent of all working days (Bergmann, 1966). On the sugar cane farms of the Philippine islands of Negros and Zebu, people generally only find sufficient work for 3·5 to 4 months a year. At peak periods, on the other hand, their labour is not enough to carry out urgent tasks in the time available. A great deal of 'work' cannot be regarded as such by the economic standards of our times; its only value is that it keeps people busy. But as the goal of economics is not just to keep labour busy but to employ it as effectively as possible, to achieve a maximum social product, countries such as these generally show great interest in systems of partial mechanisation. On the other hand a completely uncontrolled mechanisation leading to wide-scale unemployment of peasants hitherto occupied on the land could have catastrophic results. It is hoped that partial mechanisation would multiply the productivity of both human labour and the land itself. By initiating new land reclamation and settlement projects, and improving irrigation, labour demands are simultaneously increased and a growth in social product made possible.

Frequently there is a desire to set up small industries in addition; an example is the processing of agricultural byproducts. Besides processing industries, iron, steel and metalworking factories can be found, producing agricultural machinery and accessories, and fertilisers; the importance of these industries varies from country to country.

The mass of small peasant farmers can only use tractors on a cooperative basis. Mechanisation generally, therefore, has relatively little influence on animal husbandry in tropical Asia. Draught buffaloes and draught oxen are only dispensed with here and there, after partial mechanisation has been introduced, because draught animals remain essential for cultivating the smaller parcels and marginal strips.

The predominantly small-scale structure of farms, which makes the collective use of machinery necessary, also plays its part in the creation of new types of enterprise. In Malaysia, for example, jobs connected with the renewal of *Hevea* stocks (replanting jobs) are put in the hands of entrepreneurs who do not belong to the plantation company but who are usually specialists. They own tractors and other equipment themselves, and have a lot of experience in single processes such as growing the young plants, terracing, and the cultivation of so-called cover crops (mostly legumes) to avoid soil erosion and too much exposure to the sun. Often, however, it has proved dangerous for tractors, or other machinery, to be in the hands of private contractors, who would exploit the credit position of small farmers, plunging them into new obligations. As a result, farmers nowadays usually prefer to use government or cooperative tractors from a depot with several machines. Risks incurred through tractors breaking down, which were very great when the machines were privately owned, are thus reduced to a minimum. These new types of enterprise can at the same time prove useful in applying other measures which promote production, bearing in mind, for instance, the need for advice and training centres for tractor drivers, mechanics or peasant farmers generally; or the construction of a network of repair garages, which will need to be sufficiently dense and will presuppose a good road system.

So far all these prerequisites have only been met individually in most tropical countries. The changeover to mechanised agriculture only takes place gradually. Taking away part of the physical burden from the population could lead to an economic and cultural improvement. A class of trained workmen and technicians is created. The machine not only provides power but is also a catalyst for accelerated development and a status symbol for modern man. Its introduction will be an essential element in a reorientation away from the old 'society of descent' and towards the modern 'society of efficiency'.

6
Man and tropical agriculture

'The most important thing is not the soil itself, but the people living on it.'

(C. E. Kellog)

Problems of rural population

The main contribution of geography to research about developing countries is the study of existing spatial structures. In these investigations, problems connected with human group formations have a special topicality for our times. Within the framework of social geography, the attempt is being made to discover 'classes, regularities and rules of human existence in the various regions of the earth, and to illustrate the structures and processes affecting human groups which are clearly observable in those areas' (Bobek, 1966).

We can distinguish on the surface of the earth regions with only temporary settlements or none at all—regions which are, nevertheless, economically exploited—from those which are permanently inhabited and integrated into the economy. As regards their extent, these regions have undergone important changes in the course of history. In the tropics, too, economic boundaries have frequently altered. The economically significant area of the world has expanded considerably in the course of mankind's politico-economic emancipation.

It is important to recapitulate here the main facts about the earth's population (Fig. 6.1). At present this population is estimated to be over 3 600 million of whom 1 000 million live in the developed and 2 600 million in the less developed countries.[1] For the year 2000, the number of people can be forseen to be in the region of at least 7 500 millions; that means, even with a conservative forecast, a doubling of the world population between 1960 and 2000. In A.D. 2000 about 1 500 million people will be living in the developed world and 5 000 millions in the underdeveloped parts of our earth. In seven states (China, India, Indonesia, Pakistan, Brazil, Nigeria and Mexico) there are living even today 48 per cent of the world's population.

Of geographical importance is the way in which the earth's increasing population is expanding spatially; far from taking place uniformly, this expansion is producing a striking phenomenon of concentration and dispersion.[1]

[1] Estimates of world population: (Meckelein, 1966, p. 23)

	At AD 1	300 million
	1600	600 ,,
	1750	730 ,,
	1850	1 170 ,,
	1950	2 390 ,,

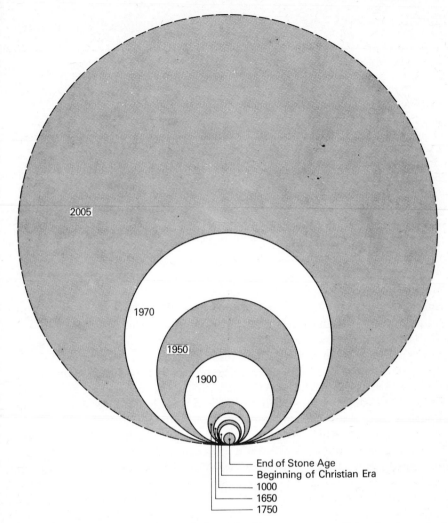

Fig. 6.1. The world's population explosion (after Dasman, 1972, and Manshard, 1973)

Alongside agricultural regions with fairly uniform high population densities, there are regions which, because of migration, have markedly changed their previous structures. Particularly striking changes take place in the areas incorporated into, or contiguous with, large tropical cities; here rural ways of life are transformed by a kind of contact metamorphism, with the compact functional area of the centre.

If we understand by 'rural population' all people living in settlements with fewer than 5 000 inhabitants, then 80 to 90 per cent of the people in the tropics belong to this category. A glance at the map of the world's population density shows that large areas of the tropics, especially in South

POPULATION INCREASE, 1900–2000 (figures in millions after Steel, 1967)

| *Developed countries** | | *Developing countries* | | |
| | | | LATIN | |
	TOTAL	ASIA†	AMERICA	AFRICA
1900 Population 554	996	813	63	120
Per cent world population 35.7	64·3	52·4	4·1	7·7
1950 Population 838	1659	1297	163	199
Per cent world population 33·6	66·4	51·9	6·5	8·0
2000 Population 1 448	5 459	4 145	651	663
Per cent world population 21·0	79·0	60·0	9·4	9·6

* Europe, USSR, Anglo-America, Australasia and Japan.
† Except Japan and the Asiatic part of the USSR.
Source. United Nations Department of Social and Economic Affairs, *The Future Growth of World Population* (Population Studies 28, 1958).

America, Africa and Australia, are very thinly populated, and in parts are almost devoid of human beings. Only in southern and eastern Asia do we find dense populations covering a large area. The average density values here are eight to ten times higher than in tropical regions outside Asia. There is a striking contrast between the relatively thinly populated tropics of Africa, Australia and Latin America, and the few densely inhabited areas of monsoon Asia, in which man had already reached a high cultural stage in early times. But even within individual 'cultural continents', and over extensive regions, there are important differences between, for example, the more densely populated forest and savanna areas of West Africa, and the thinly populated 'middle belt' which lies between. Another striking contrast is that between the scarcely populated forests of the South American lowlands and the concentrations of population in the Andean highlands. This is a case where impeding physical geographical factors made settlement more difficult. Without falling prey to a deterministic approach, which offers itself enticingly, especially in the study of the tropics, reference must be made, nonetheless, to the often close dependence and reciprocal reactions between climate, vegetation, soil and man in the *cultura agri*. In the past, the more or less direct dependence of man on climate in the tropics was often pointed out; nowadays we should accept this statement with greater caution, especially in view of the great population density and high cultural standards in parts of tropical Asia. The development of soil and vegetation, itself closely connected with climatic factors and with numerous human, plant and animal diseases, had a great influence on the ways and means by which man dominated his environment and contributed to the development of various ways of life and types of economy.

The reasons for the presentday distribution of population in the tropics have not been studied thoroughly. Apart from incomplete statistical material, there are only a few studies which deal with questions of population density in its relationship with natural, cultural, historical and economic factors.

Besides such well-known concentrations as India and China, the population concentrations in parts of Nigeria, the interlake area of East Africa, and southeast Africa are particularly striking. In Southeast Asia, Java and the Mekong delta have a high population, whereas New Guinea, Borneo and Burma are relatively thinly populated. Also striking is the settlement pattern of Australia and South America, spreading from the coasts or from the major rivers (e.g. Parana–Paraguay).

Looking at a population map of the tropics (Fig. 6.2) there is one phenomenon which immediately catches the eye: the African tropics,

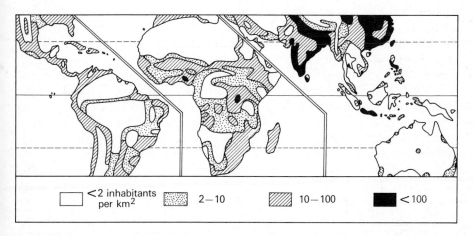

FIG. 6.2. Simplified actual population densities in the tropics (after Daveau and Ribeiro, 1973)

taken as a whole, are relatively thinly populated, the Asiatic tropics much more densely. In monsoon Asia, population concentrations are mainly found in the alluvial plains of the large lowland rivers. In Africa it is precisely these areas which have remained more thinly inhabited up to the present, while in contrast, the more densely populated areas lie in the highlands (e.g. Rwanda, Burundi, Malawi). In Africa the slave trade was a specially negative factor affecting the distribution of population. In more densely populated tropical Asia there have always been more intensive cultural exchanges with the well-developed subtropical cultures to east and west, so that tropical Asia has enjoyed a higher civilisation and better techniques.

The development of agriculture in the tropics shows clearly that the tropical environment has frequently forced man to react in a certain way.

As is well known, Toynbee (1934) tried to reconstruct the development of mankind in a dialectical rhythm of 'challenge' and 'response'. The 'challenge', which Toynbee would interpret not simply as natural conditions, was so overwhelming in some areas that man had to succumb to natural environmental conditions; thus, no higher culture developed, whereas, on the other hand, such cultures were created by man as a 'response' in more favourable areas. This dialectical thesis seems somewhat speculative and rather simplistic. Moreover, it was based on the principle of relativity and thus in our context has little predicative value; there is no 'absolute challenge'. In the relationship between man and his environment, it essentially depends on the group whether there is a 'challenge' or not. Another question is how the 'challenge' is met. For instance, there is no necessity at all to respond to any given challenge by developing a culture. This relationship can be manipulated at will, therefore, and can be applied to the whole earth in its relativity and universality.

Allan (1965) introduced the concept of 'critical population density'. By this he means the maximum population density which is possible with a certain level of cultivation without causing soil deterioration. Although this concept was first applied to simple subsistence economies in central Africa, it is valid also for other types of economy in which the 'capital' of the reserves provided by nature is quickly used up, and the future existence of the ever-increasing population depends on the steadily decreasing 'interest'.

It is true that a pronounced dominance of biotic and climatic factors is certainly not typical of all parts of the humid tropics. In the densely populated Yoruba and Ibo districts of Nigeria, for instance, the critical population density seems to be very different from that in Allan's main research areas in central and East Africa.

Obviously, man first of all took hold of the unexploited resources of his environment; in the case of the tropics, the sparsely inhabited forest and savanna regions in particular. It is only when almost all these areas have been occupied that the population reaches a critical borderline, at which stage it has to decide whether to continue exploitation, causing perhaps irreparable damage, or to change over to new forms of production.

Using archaeological and botanical research, McNeish (1964) illustrated the way the economy had developed in the Tehuacan valley in Mexico. He studied the effects, first of all, of rainy season cultivation; then of the later irrigation practice; and finally of the development of crafts and trade, on the population-carrying capacity of the valley. The latter has increased 5 000 times within the last 8 000 years, from the individual family to organised community life. Similar figures can be provided for many places to show the effects of development from hunting and food-collecting groups, through peasant shifting cultivation and land rotation, to permanent cultivation (Hutchinson, 1966).

For other parts of the tropics, too, certain threshold values have been roughly calculated for the population-carrying capacity of the land, during the period of transition from shifting cultivation and land rotation to more permanent forms of cultivation.

On the assumption that an average family of five persons could live on food provided from one hectare, which under a system of shifting cultivation would be cultivated every ten years, van Beukerling (Pelzer 1948) estimated a critical threshold of about fifty inhabitants per square kilometre on the Outer Islands of Indonesia. For south Brazil Waibel (1955) proposed a critical limit (family size 5 to 7; 5 ha under cultivation; 10 to 12 years fallow) of 5 to 13 inhabitants per sq km. Using other indirect methods for a region east of Belém, Valverde and Dias (1967) deduced the nonpredatory limit of shifting cultivation to be about nine inhabitants per square kilometre. Once these critical points are surpassed the fallows have to be reduced and a progressive degradation of the environment (especially of soil and vegetation) sets in.

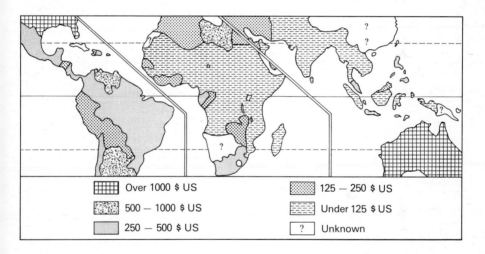

FIG. 6.3. Gross national product per capita (1963) (after Daveau and Ribeiro 1973, cf. also Fig. 6.2; World Bank 1972)

However, because of the numerous ecological, social and economic variables, these figures have to be accepted with caution. Under land rotation in the forest areas of southern Ghana, a critical limit seems to lie at about sixty inhabitants per square kilometre (Boateng, 1962, p. 17). For the densely populated areas around Kano or Sokoto in Northern Nigeria, Prothero (1957, 1958) and Mortimore (1967) have quoted values of 100 to 160 people per square kilometre. It must be taken into account that in the vicinity of towns it is often difficult to make allowance for additional income from crafts and trade, and also that irrigated cultivation (*fadama*) already plays a part in farming during the dry season.

Another very controversial question is the population potential of a

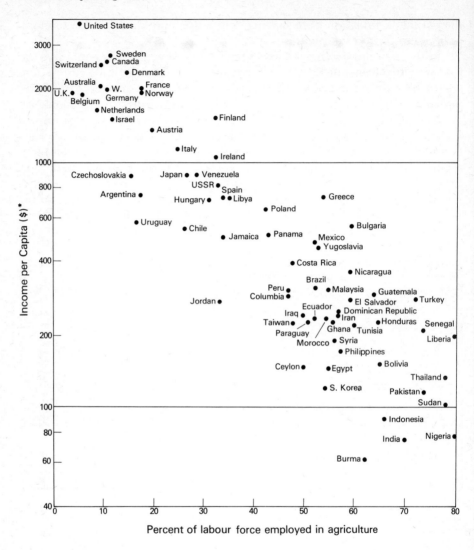

FIG. 6.4. Income per capita, 1967, and percentage of labour force in agriculture, 1965, developed and less developed countries

region such as the Amazon Basin. At present, this population of about 5·5 million people on 5 million square kilometres is very thinly spread—about one inhabitant per square kilometre. The optimistic estimates of Penck (1924, 1941), Fischer (1925) and Hollstein (1937) for a potential population of several hundred million, seems at present further from reality than ever. These estimates were based on several wrong assumptions without the better founded knowledge of this region that we now possess. In this connection the studies of Brücher (1968), Monheim (1965), Sioli (1969), Valverde (1968), Wilhelmy (1970), Glaser (1971), deserve further attention. It is only after taking into account regional differences of various greater units like *terra firma*, *varzea* and the slope region of the Andes, that any guesses on the demographic and economic potential can be attempted.

Geographers (Bobek 1961; Uhlig 1962) have frequently stressed the obvious fact that carrying capacity is not only a problem of physical geography but that it also has, to a very large extent, a social geographical aspect. Thus, as well as basic demographic facts, the geographer pays most attention to questions of the structure, the flexibility and the carrying capacity of economic development areas (e.g. in regions of increasing or decreasing population).

In order to be able to interpret the spatial patterns of agricultural regions it is necessary to study the very intricate natural, social and economic structures operating on them. After first dealing with the importance of diseases, a few typical examples from the great number of tropical agricultural regions will be selected and looked at from different points of view. In these short sketches, which in part are the results of previous studies by the author, the stress has purposely been put on different factors in order to present different starting points.

The importance of diseases in the agricultural regions of the tropics

The domestic agricultural situation in tropical countries to a large extent defines the state of the inhabitants' diet (see p. 32) and is partly responsible for the unbalanced diets or even food shortage among tropical peoples. Deficiency diseases are as common as other illnesses (such as smallpox, typhoid or tuberculosis) which spread disastrously because of bad hygienic conditions and generally low standards of living in the tropics.

In many arid and semi-arid tropical regions, food availability depends largely on the yields of one or two cereal crops at the end of the rainy season. In years with insufficient rainfall, or when the rain comes at an unfavourable time, or when swarms of locusts have ruined the seed or the harvest, the amount of food available is not usually enough to cover demand until the next harvest. If the loss is exceedingly high, or if two bad harvests succeed one another, the food shortage can be so serious that real famines may occur. Scarcity in the period preceding the new harvest is actually the rule in many regions (the hungry season), but it is especially harmful to health and efficiency at this particular time, when the people should be in full possession of their strength for the strenuous work of the harvest.

A particular disadvantage is the insufficiency of highly valuable animal protein. Soon after children have been weaned there appear the first symptoms of the illness caused by lack of protein, known by its African name of kwashiorkor. It can especially be seen when, although the amount of calories in the diet is sufficient, not enough protein can be provided. Its clinical symptoms are growth disorders, emaciation, deposits of liquid in the tissue, especially of the legs, and a bloating of the body owing to the liver swelling and water accumulating. Even among children who do not show any external symptoms of illness, infectious diseases such as measles, which for us are uncomplicated, are usually fatal because of the lack of protein.[1] Mapping the disease has helped to make it clear that kwashiorkor mainly occurs in regions where children are brought up on a basic diet of cassava (manioc) and bananas. The disease is rarely observed, on the other hand, in areas with a diet of cereals, such as millet. These differences can be explained by the varying protein content of these plants.

The nutritive content of the diet clearly depends on the various techniques used in food preparation. But it is not only old-fashioned methods which can endanger tropical peoples through their diets. The development of new techniques, too, can be dangerous for them if they are not supervised. Beri-beri, for instance, only spread so devastatingly in East Asia after rice began to be supplied by mills where Vitamin B_1 was lost in the dehusking and polishing process. In some areas the consumption of animal protein is still artificially prevented by taboos. In parts of Tanzania, for example, women and children are forbidden to eat hens' eggs; in some regions of Uganda this prohibition applies to men. The diet of many tropical peoples is generally poor in minerals as well; thus, the craving for salt is very great, which explains many cases of 'geophagy',[2] such as have been observed in North Borneo, Vietnam and Africa.

As well as these general deficiency diseases, specific tropical diseases have very serious consequences for the population. It is not uncommon for debilitation and incapacity to carry out manual or mental work to be the result of these latter, thus completing the vicious circle. Thanks to Rodenwaldt and Jusatz (*Atlas of World Diseases*, 1952–1961) the limits to Roden- of acute and chronic diseases have been shown, and the routes taken by great epidemics suggested. Equipped with this exact cartographic delimitation of the distribution of illnesses, medical and ecological analysis could be promoted and efficient measures to combat the diseases taken. It

---

[1] The situation becomes even more acute as general resistance is lowered; this is clearly expressed in the high death rate among children of one to five years.

[2] In northern Ghana a drink is prepared with water which is so rich in kaolin that it looks like milk. In Honduras a drink is brewed by soaking maize in lime water. What has not been clarified yet is whether geophagy (the consumption of certain loamy or clayey soils) allows the intake of a kind of 'drug', as in the chewing of betel and lime, or whether it is done for medical reasons. Among American Negresses the eating of laundry starch was common, especially during pregnancy.

became obvious that many diseases which do not require intermediate carriers to be infectious to man have spread worldwide and expand particularly along traffic routes. In all the other cases, however, where intermediate hosts are involved, the spread of these diseases depends on the living conditions of the carriers (insects, and rodents, among other mammals). The mobility of the carriers thus depends on geographic, mostly climatic, factors. In yet other cases, the spread of the disease depends on the ability of the virus to survive in the open air, outside the organism which it attacks (also bacteria spores and the eggs of worms living in the intestines of men and animals).

The most widespread tropical disease is malaria. It also occurs in swampy areas in the subtropics and in those regions of the temperate zone where there is a mean temperature of over 16°C for several months. Deltas and lagoons on tropical coasts, flooded areas (e.g. rice-growing regions) and swamp forests are dangerous seats of malaria. In tropical West Africa, South America and South Asia many of the inhabitants are infected by malaria. Weakened by a deficient diet, they often show but little resistance to infection. Thus, a famine in Ceylon (Sri Lanka) caused by the non-arrival of the monsoon, and its corollaries of drought and harvest failure, was followed by a malaria epidemic. According to one estimate, malaria is reckoned to have been the direct cause of death of 30 million people in India between 1901 and 1931. Moreover, it prepares the way for the spread of other diseases and reduces the convalescents' ability to work.

The larvae of mosquitoes need certain environmental conditions in water for their development. These conditions vary according to type: shaded opaque water, or slowly running water which has been warmed by the sun. Thus, in almost all the warm, humid tropic belt the anopheles mosquito finds conditions which favour rapid propagation. Nevertheless, it has been proved possible to prevent the breeding of anopheles mosquitoes on both a local and a regional scale in the tropics, and to clear areas of the epidemic. The Tonkin delta will serve as an example (Gourou, 1953). Whereas there are hardly any cases of malaria in the delta, the surrounding mountains are very unhealthy. The inhabitants are fully aware of this; labourers on the farms at the edge of the delta return every night to the villages of the plain. The Tonkin delta has one of the highest densities of rural population in tropical Asia. It is almost entirely used for wet rice cultivation. Two or even three harvests per year are possible on 50 per cent of the cultivated area of the region. In contrast, the surrounding mountainous land is thinly populated and the cultivated area, which does not comprise more than 5 per cent of the total, never yields more than one harvest per year. It is surprising, at first glance, that living conditions in the delta are so healthy, since the land is covered by flooded ricefields and pools. The main channels of the Red River and its affluents, together with other narrow meandering ribbons of water, could theoretically turn into a breeding ground for anopheles mosquitoes during the dry season. The farmer, however, cultivates all the land which is uncovered when the water-level is low, and has built a network of dykes, drainage and irrigation systems. Accordingly, the health of the inhabitants does not depend on natural factors but is the result of man cultivating the land and controlling the water. The agricultural methods and most of the social and political forms of organisation which have enabled the farmer of the Tonkin delta to make these achievements can be traced to Chinese models. They did not originate in the tropics but were applied by the Annamites in the tropical Tonkin delta, which they opened up for rice cultivation and freed from malaria by cultural-technical measures.

Whereas malaria is very common in almost all tropical countries, sleeping sickness, trypanosomiasis, is characteristic of tropical Africa. It was first observed in Sierra Leone. At the end of the nineteenth century, and especially in the twentieth, it spread from its endemic area—along the river courses of tropical West Africa, Senegal, Niger and the Congo basin—to Angola and central Africa as far as Lake Victoria, mainly because of population migrations. At the turn of the century, the greatest sleeping-sickness epidemic was located in East Africa, during which two-thirds of the population died in parts of Uganda.

The trypanosomes are transmitted to man by the tsetse fly, especially *Glossina palpalis* and *Glossina morsitans*. The tsetse fly finds the special conditions of temperature and humidity which it requires in the shelter of the uncleared forest. Bringing an area completely under cultivation and cutting down even low bushes therefore destroys the breeding grounds of the *glossina*. It is only possible, however, to make an area healthy if the population is large enough to utilise the land completely. A high population density, therefore, is a necessary precondition rather than a consequence of human health in such regions. Gillman (1936) showed that in Tanganyika it was not so much the tsetse fly which had driven the population out of many areas but rather that man had simply left the land to it without resistance. Other kinds of trypanosomes cause nagana in cattle and other domestic animals in Africa. The danger is so great in the fly belt that animal husbandry is severely restricted.

Yellow fever is another dangerous tropical disease. Both the viruses and the gnats transmitting them (*Aëdes aegypti*) need high temperatures—a mean minimum air temperature of at least 21°C. In Africa south of the Sahara yellow fever is not confined simply to active endemic centres on the Gulf of Guinea but also occurs in a latent form as far as southern Ethiopia. In tropical America it spread inland from the West Indies and the east coast. Towns for the most part have been rendered healthy by destroying larvae in all stagnant waters.

Parasitic diseases are no less weakening to man. To this group belong amoebic dysentery and numerous worm diseases. What is needed, above all, is an improvement in sanitary conditions, which would prevent the dissemination of disease to a large extent. Hookworm (*Ancylostoma*) must be mentioned as one of the most important worm diseases. It is true that it is not fatal, but it often leads to physical and mental lethargy and also to anaemia, because of constant loss of blood. Bilharzia has once more become a frequent worm infection in the tropics of Africa, South and Central America, and in Egypt. It is caused by parasitic schistosoma which are transmitted by snails to human beings when bathing in the open air or working on irrigated land. Both widespread amoebic dysentery and the various worm diseases can only be combated successfully if the population is systematically educated about health. Since insects are the main disease-propagating agents, one of the most important tasks of a health service in the tropics is the spreading of insecticides. However, in order to fight them effectively, the ways of life and ecological relationships of insects must be precisely studied.

One main problem in fighting insects is to bring the insecticides as close to them as possible. To bring poison into direct contact with harmful insects is usually made difficult in tropical wet forests and commercial monocultures because of the way the plants grow. One has only to think of the height to which oil palms grow and the rich foliage of tropical banana plantations, or the way in which the tsetse-fly larvae are buried for protection in the soil. Measures for clearing the terrain are more effective than the chemical treatment of breeding places, or the use of insecticides which, moreover, represent a permanent financial burden and a growing environmental danger.

In Sumatra, extensive territories in the Tarutung valley and the Pahai plain were made healthy by splitting up watercourses and thus diminishing the velocity of the rivers. The work was supplemented by embankments built across the watercourses, which raised the level of the land. The channels were joined up again later. The watercourses were thus improved in both the speed and the force of the current. In several regions, not only was malaria stamped out in this way, but valleys were made cultivable. There is no general scheme which can be applied to ensure the successful sanitation of an area. On the contrary, the exploitation of coastal mangrove forests for timber, for instance, can turn a healthy coast into a centre of disease through unintentionally causing silting and the formation of pools of brackish water. Whereas deforestation implies danger in this case, it is of course necessary in areas where sleeping sickness is being fought, as in the fly belt, on the shores of lakes or in regions of riverine forest.

Before sanitation measures can be launched the geomorphological, climatological and hydrological situations, and their biological connections, have to be analysed. Only a geomedical examination of all factors can indicate the ways in which man may be able to impede the beginning or change the course of influences which endanger life. It is still true of many parts of the tropics that the towns are the healthiest places; here, contact with nature has already been broken and hygienic conditions are often better than in the countryside.

In order to understand agricultural structures in the tropics, sufficient attention must be paid to the situation as regards disease, which has only been lightly outlined here. Agricultural geographers studying the tropics have so far mainly confined their attention to the production and distribution of foodstuffs. The analysis of food consumption and the state of health connected with it has been neglected. The same applies to the relationship between health and density of population. Whereas at one time tropical diseases prevented too much growth in population because of the high mortality rate, nowadays problems of overpopulation are getting more and more acute. An improvement in the disease situation thus often goes hand in hand with a worsening of the food situation. It is important, first by studying comparable individual cases, to find out all details of the relationships between agrarian structure, food consumption, population density and health or disease conditions in selected agricultural regions.

The development of agriculture in areas settled under European influence in Latin America

It was in the fifteenth century that Western civilisation began to show a greater interest in the hitherto unknown world, whose 'unveiling' had been completed, at unprecedented speed, by the end of the nineteenth century. Throughout this period Europeans exerted their influence in tropical countries, a process often referred to as 'Europeanisation', and overseas colonial empires were created. One of the most important consequences of the process was the ever-growing demand for tropical products, which soon became indispensable raw materials for the European countries. On the other hand, overseas areas became more and more important as markets for readymade goods. In the first few centuries there were strong economic and cultural links between the colonies and the mother country, which were intensified to some extent by waves of emigration. In spite of the fact that there has been a trend towards the 'de-Europeanising' of the world since the beginning of the twentieth century, the European and American civilisation and economy have not ceased to make their imprint on all the earth's peoples. World economic interdependence has become ever tighter.

The challenge of the tropics to the white man, and his biological and social acclimatisation, in the broadest sense, was called 'tropicalisation' by the Brazilian sociologist Freyre. Brazil, where European, Amerindian and Negro races mixed, is a good example of this phenomenon. The result was a Mediterranean-Latin-Christian civilisation without any strict racial separation.

In the beginning the Portuguese element was very important for the development of agriculture in Brazil. The first Portuguese settlers, who came from overpopulated Portugal and the Azores at the beginning of the seventeenth century, adopted the Amerindian *roça* system of subsistence agriculture. An aristocratic-feudal social system helped the introduction of a plantation economy; sugar cane plantations were created which were

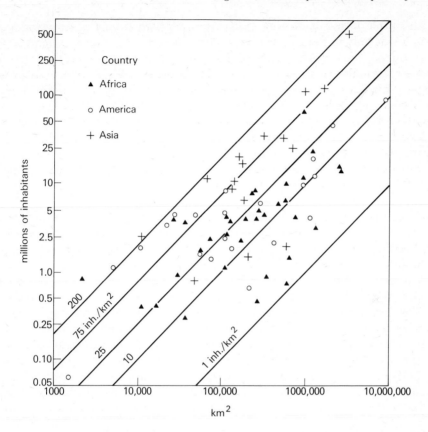

FIG. 6.5. Area and population of independent tropical countries, 1967 (after Daveau and Ribeiro 1973, and *Statistical Yearbooks* of the United Nations)

worked by slaves. Later, animal husbandry and cocoa cultivation developed over extensive areas, and rubber-collecting and finally widespread cultivation of coffee were carried on. The existing cultural landscape was still further diversified by the immigration of Italian, German, Polish, Lebanese and Japanese ethnic groups.

Great mobility among this immigrant population is a special characteristic up to our own day. As in North America, the immigrants, obsessed by overwhelming optimism, wandered from one part of the country to the other, trying their luck. Local agricultural developments were favoured by the discovery of diamonds, gold, iron ore and precious stones, and by the increasing attraction of urban and industrial centres.

Waibel (1955) and Pfeifer (1953) have described in detail the problems of German farmers in the tropical surroundings of southeast Brazil: 'In the last hundred to two hundred years colonists have gone through the same phases of development as European agriculture over a span of one to two thousand years' (Waibel). At first, permanent settlement was combined with simple forms of shifting cultivation adopted from the Amerindians, and it was only in the course of time that better land rotation systems were introduced. Small local centres came into existence with facilities for processing agricultural products (manioc). Besides native crops (maize and beans) European crops were raised, and cattle breeding was developed alongside pig keeping. Only about 5 per cent of the colonists actually reached the third phase of development, in which the stage of land rotation was followed by true crop rotation with ploughing and manuring, and in addition cash crops were cultivated.

For Central Brazil, Glaser (1971) has given an account of new agricultural developments.

Until recently the colonisation of Central Brazil has been characterised by a separate development in the areas under forest and in the campos. The land use in the campos was restricted to extensive cattle-rearing on large estates, whereas in the areas under forest shifting cultivation, mostly carried out by small farm-holders, dominated. This situation has completely changed over the last ten years as far as the new pioneer zones of the forest regions are concerned. There today cattle-raising fazendas play the most important rôle in the process of colonisation. The impulse for this new development has been given by the building of some important new roads connecting for the first time by proper means of transportation many parts of Central Brazil with the Brazilian centres of consumption near the coast. New possibilities for market production arose, while the freight rates and the time necessary for transportation decreased considerably.

Because of transport in special modern trucks cattle no longer loses much of its weight as it did before on its long marches (*boiadas*) to the slaughter houses of São Paulo. Today livestock-rearing gives higher profits, which enable the stock farmer to intensify his way of farming. He has become interested in introducing artificial pastures, as one head of cattle needs 3–8 ha of natural campos pastures but only one ha of artificial pasture. On the other hand it has proved far more difficult to turn the poor soils of the campos region into artificial pastures than the fertile soils of the areas under forest.

Thus the untouched forests in Central Brazil have become very attractive to the expansion of cattle-raising. After having slashed and burned down the forests and before turning them into pasture the farmers grow rice, maize, beans and manioc by means of the primitive roça-system for two or three years. By taking advantage of the high yields per hectare on freshly claimed forest soils they lower their costs of investment.

However, it has become a great problem to find enough people to do the clearing and the cultivating during the first years. Central Brazil suffers from a severe shortage of manpower. Even many of the numerous immigrants from other parts of Brazil rather prefer to remain under miserable circumstances in the urban regions than to move into the country because of the even worse working and living conditions there. Rather curious practises of hiring labourers have been developed.

Another very serious problem arises from the clash between the wealthy stock farmers who claim a sort of land monopoly in the forest regions and the poor squatters. Very often these conflicts are decided by weapons.

Considering these facts to a certain degree the hopes have not been fulfilled that the 'March to the West' might bring some relief to the social problems of Brazil. On the contrary, the poor immigrant to Central Brazil encounters strong hindrances, if he wants to acquire land of his own. In our days also the frontier towards the vast forests has become decisively characterised by large estates.

Even the integration of this region with the economy of Brazil has not yet been solved satisfactorily. European immigration is no longer important. Only the growing number of *mestizos*, of about one million per year, could provide a source of settlers. Many of them who worked during the earlier rubber booms in Amazonia have since left, particularly for the cities. Some Japanese settlers have been successful (Tanaka 1957), introducing pepper and jute cultivation, and vegetable growing reaching a high intensity, so that they set an example for the caboclo cultivators. A careful and detailed analysis of the Japanese settlements in Amazonia may throw some light on the potential of the whole region.

An early socio-economic polarity, between large landowners and slaves, was typical of Brazil. Family ties were, however, strongly stressed. In spite of racial and social barriers, 'transgressions' were frequent, however. The merging of three races into a European tropical civilisation in Brazil can be viewed as an interesting model for the racial history of mankind. Without the 'sweat of Africa', says Freyre (1965), the Brazilian synthesis would not have succeeded.

In contrast to Brazil, the former Portuguese colony, which was opened up to settlement mainly from the coast, in the former Spanish colonies the highlands were in the forefront of development, since they were healthier for Europeans (e.g. Ecuador, pp. 149–52).

Sandner (1961) has described the agricultural structure of Costa Rica, with its predominantly white population who live in the cooler zones of the *tierra templada* and the *tierra fria*, at heights of over 1 000 metres. A larger number of settlers has already penetrated from the central upland basin into the tropical lowlands along the Pacific and Atlantic coasts. In so doing, the white settlers of Costa Rica have had the advantage of not encountering a numerically strong native population. This disorganised pioneer colonisation of the lowlands, which was mainly carried out by the peasant population already living in Costa Rica, began about 100 years ago. Among the main reasons behind it were the social tensions on the land in the densely populated core areas: on the one hand the increasing fragmentation of peasant holdings and on the other the process of property concentration, which started because, through getting into debt or because their holdings were uneconomic in size, the peasants were forced to work as labourers to earn additional income. In order to escape from this dependence they sold their land and migrated to the wild areas being newly colonised.

Members of all social groups took part in this disorganised colonisation: poor squatters and agricultural labourers, speculators and small-scale farmers, large landowners and entrepreneurs who wanted to invest their capital in land, livestock or forest. Of particular importance was *parasitismo* —the unlawful occupation of private or public land by squatters, who came to occupy more than a third of the land in many marginal areas of Costa Rica. According to 'squatters' rights', which already had legal validity in the nineteenth century, anybody could acquire land as his own lawful property if he could prove that he had cultivated and improved it so that its value had increased at least threefold.

The same social problems appeared again and again, however, in the newly cleared areas (see Fig. 6.6). Lack of capital, abundant land and isolation encouraged extensive farming and fostered a tendency to subsistence farming. Road developments usually came decades after colonisation, and even then only due to the initiative of the colonists. Sometimes these developments brought penetration by large landowners, and the poor population migrated. In cases where the cleared areas were far away from markets, the owners of stores (*pulperías*) could acquire considerable extents of land by granting credits and advances on harvests, and then getting the poorer colonists into debt. Such impoverished settlers then either started work as *peones* (agricultural labourers) or tenants on the large and medium-sized estates which were gradually invading the area, or they sold their farms (which they had obtained for nothing) and followed the advance of the pioneer fringe. Such developments in the marginal regions threw doubt on the process of pioneer colonisation, not only from the economic point of view, because it was unremunerative in many cases, but also from the social aspect, because the primary aim had been the unification of a shifting population. Since the setting up of the Instituto de Tierras y Colonización (ITCO) in 1963 a certain change has taken place. The illegal occupation of land by squatters has been forbidden. Interested candidates are put on a waiting list and are allotted land in the new agricultural colonies which are being set up with such speed.

In the colonisation zones of Costa Rica the structures of the settlements vary considerably from, for example, those in the plantation territories of the American United Fruit Company. In Costa Rica the primary settlement type is the isolated farm with its lands in a single block adjoining it. There are not usually any other farm buildings since the only tools used are the digging-stick, bush-knife, shovel and axe; any kind of stable is unknown. The 'farm' thus consists of a living hut, a pig pen, sometimes a

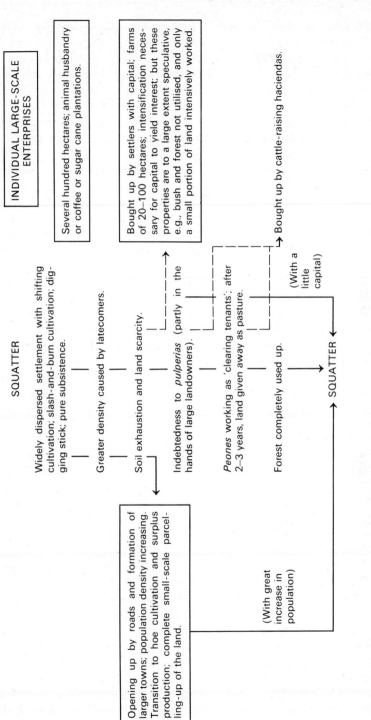

FIG. 6.6. The development of the areas of uncoordinated agricultural colonisation in Costa Rica (from data in Sandner's studies)

small sugar cane press and a shed for harvesting tools, saddle and harness. Rural settlement centres only gradually emerge with an increase in population density and the development of a communication system.

As one goes from core market areas towards the periphery, the spatial agricultural structure of Costa Rica shows a gradient from more intensive to more extensive types of economy. The contrast between the central densely populated highlands and the flat sparsely populated coastal lowlands is also reflected in population density. Waibel already referred to this centre periphery order among types of economy as early as 1948. In the Valle Central he distinguished the following zones of agricultural systems, arranged in circles: (1) monoculture of coffee; (2) coffee and sugar cane; (3) meadows and fields; (4) forest-field rotation; (5) Animal husbandry. Starting points for the formation of Thünen's Rings exist, but not in the marginal areas, where they could not develop because of the predominance of subsistence farming without any market connections (Sandner, 1961, 1962, 1964).

In the course of colonisation various spatial types emerged in tropical America. Pioneer fringes progressed on a broad front and compact cultural landscapes developed. But the frontier (*frontera*) was also pushed forward in a series of jumps, and there were new starting-points here and there. Apart from the type of economy practised and the distribution of population, other reasons for the great variety to be seen in the direction of development are communications possibilities, depending on relief and other features of the physical environment (Fig. 6.7).

In his work on the vegetation, land use, and agricultural potential of El Salvador, Lauer (1956) presented an interesting map in which he calculated population densities within vegetational units. He shows that these units have very different land use potentials and that for their rational future use more detailed 'ecological planning' is necessary. While some regions (e.g. the evergreen lowland forest) lend themselves to more intensive forms of arable farming or even mixed agriculture, others (e.g. the Chaparral) urgently require reafforestation.

For some tropical lowland areas of Central America, an instability of tropical land use is typical. Stouse (1970) has investigated this for the Atlantic lowlands of Costa Rica where bananas are being produced for export. A number of historical and areal factors, affecting the plantation system of this region have contributed to this unstable situation.

In recent decades a few refugee peasant settlements have also been founded in the American tropics. Since 1958 groups of Mennonites, originating in Russia, have settled in the northern parts of British Honduras, being one of the first of such groups in a tropical lowland. The colonists came from the dry regions of northern Mexico, where they gave up their properties because of difficulties with the government. They receive state aid to help them to revive the neglected agriculture of the country. They have complete religious freedom and they were allowed to administer their

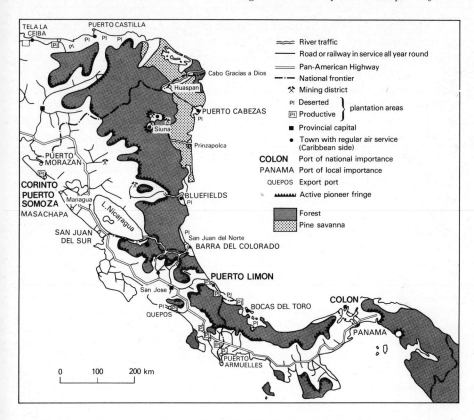

FIG. 6.7. Pioneer fringes in the Caribbean forest region of southern Central America (after Sandner, 1964)

settlements themselves. With government help larger properties were bought, on to which they introduced dairying and poultry farming. Nowadays, British Honduras is much better provided with poultry than her neighbours. However, problems arose out of the individualistic attitudes of the Mennonite groups.

There are many examples of European colonisation among the islands of Central America. Cuba and Puerto Rico have populations in which the strong white element is mixed up with other ethnic groups. Most of the Europeans came as peasant immigrants from Mediterranean countries. As colonists, for the most part they attained a living standard equivalent to that in their homelands. On both these West Indian islands there were special circumstances in the favourable climatic conditions and the economic and political situations, arising, for example, from the United States's obligation to purchase Cuban sugar and from the long-lasting colonial status of Puerto Rico.

The different effects of distinct racial groups on the culture and agriculture can be well observed if one compares the Negro republic of Haiti with the Dominican Republic, which has been more under white influence. The frontier between the two states, both of which are on the island of Haiti, was unstable for over a century. Following on the military invasions by Haiti of the Dominican Republic, which took place from the middle of the nineteenth century onwards, there was a great influx of Negro settlers from the former into the latter. The old Dominican agricultural structure disintegrated. Traffic and trade were redirected towards Haiti. After the treaty of 1936 this process was reversed and the Dominican government took strict measures to tie the border provinces more closely to itself, in order to 'Dominicanise' them both culturally and economically.

The transformation of the Caribbean islands under British rule into important sugar-producing agricultural colonies took place in the seventeenth century, in the short span of only fifty years. This rapid development is all the more striking as neither the British planters nor the African slaves had any experience of sugar cane cultivation. One explanation can probably be found in the immigration into Barbados, Jamaica and other Antillean islands of Portuguese Jews who fled from Brazil to escape persecution in the middle of the seventeenth century. For many decades in the seventeenth and eighteenth centuries, at a time when Jewish refugees could not obtain permission to immigrate into England itself, they played a considerable role as innovators in the British colonies in Central America.

Although during the colonial period European influence generally made itself felt in various distinct spots and from the periphery (the coastlands and islands), later on it spread increasingly over the whole area. Besides the disputes between individual European powers it was the development of transport, from the sailing ship to the jet aeroplane, which was the real pacemaker in the opening up of the tropics. New techniques, especially of preservation and refrigeration, made it possible to transport fruit and processed raw materials through the humid torrid zone.

The European colonisation of the American tropics, in particular, is closely associated with problems of acclimatisation. These problems include not only the adjustment of the body to different climatic conditions but also, which is very important, the overcoming of the socio-economic and cultural tensions which confront man when he stays for any length of time in an alien environment. If the economic situation is unfavourable and previous aspirations can no longer be fulfilled, 'adaptation to lower standards' can take place (Sapper, 1923). Acclimatisation is still extremely topical, in spite of the 'decolonisation of the world', because owing to modern means of transport and presentday worldwide economic exchanges, many people of all races are forced to acclimatise themselves for short periods.

Sierra and *Costa* in Ecuador

In tropical mountain areas the altitudinal limits of agricultural zones are often defined by the decrease in temperature with altitude. In the Andes of Central and South America the following belts are distinguished:
1. the hot *tierra caliente*,
2. the warm-temperate *tierra templada* (roughly 1 000 to 2 200 m) ;
3. the cool *tierra fría* (10°C mean annual temperature; up to about 3300 m) ;
4. the *tierra helada*, above the limit of daily alternating frosts (see Fig. 4.4).

Such a variety of different habitats and production areas fundamentally influences the possibilities of economic development for tropical mountain regions. The geographer is interested in the reciprocal effects and means of exchange between these agricultural regions which, incidentally, have been evaluated rather differently by man in the course of history. A good example of the changes in these relationships is Ecuador, whose areal structure is mainly made up of two large natural regions, the coastal fore-land and the mountains. When Alexander von Humboldt travelled here at the beginning of the nineteenth century, 90 per cent of the inhabitants lived in the *sierra* and barely 10 per cent on the *costa*. Between that time and the present day, the proportions have changed to 58 per cent for the *sierra* and 40 per cent for the *costa* (Sick, 1963). The remaining 2 per cent live in the very thinly populated wooded lowlands of Oriente province, which lies east of the Andes and which physically belongs to the Amazon basin.

The migration of population from the mountains into the western low-lands bordering the Pacific has been going on right up to the present. In the pre-European period and during Spanish colonial times the high valleys and basins offered better possibilities, mainly for climatic reasons. The highlands were the core area, economically and politically, both for the old Andean states and for the Spanish colonial power. Agriculture and mining were concentrated here, whereas there were only a few isolated settlements on the *costa*. With a shift of mining to the south greater agricultural activity began to develop in the lowlands, from the middle of the seventeenth century onwards. Besides the gathering of food and other commodities in the forests (Peruvian bark, balsa wood) the systematic cultivation of the cocoa tree began in the eighteenth century; cocoa is native to South America. Alongside this crop, animal husbandry and the raising of tobacco and rice began to be important in the drier southwestern areas of the *costa*.

In the higher regions of the *sierra* both the population and the economy have retained their conservative features up to the very recent past. It is true that the hoe has replaced the Indian digging-stick, but the old 'hook-plough' is still used. Mechanisation has only made progress to a small extent, partly in association with irrigation, in some inter-Andean valleys and basins.

Social conditions have not greatly changed either. Alongside the big

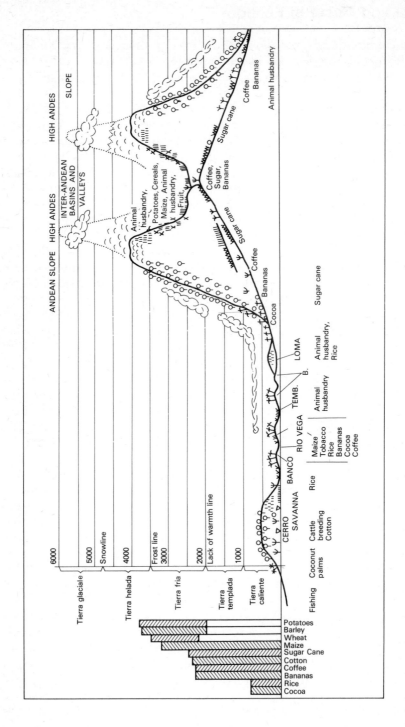

Fig. 6.8 Ecuador: vertical land use profile

latifundios, which sometimes go back to the days of the *conquistadores*, there is a large number of *minifundios* belonging to the Indian population. Even elements of the old colonial organisation (*encomienda* system) still survive in altered forms in the *huasipungo* system, according to which the Indians put in a few days' work on the estates of the white or *mestizo* landowners, in return for a small payment. Whereas the Indian population on the coast has achieved a higher living standard because of the shortage of labour in the region, a surplus of labour has prevented any change in old-established conditions in the *sierra*.

The systematic opening-up and settlement of the western foreland of Ecuador took place in the nineteenth and twentieth centuries. Modern communications development and state planning facilitated this expansion. Since the eighteenth century important changes have taken place within the agricultural regions of the *costa*. Cocoa occupied a dominant position for more than two centuries (in 1900 still accounting for 20 per cent of world production; today only 3 per cent); but a catastrophic regression of cocoa production occurred owing to plant diseases. The trend towards tropical monoculture still continued. A changeover to banana cultivation took place, to satisfy the North American market; this also contributed very much to the expansion of cultivated land in the *costa*. Coffee is grown in the central coastal area north of Guayaquil, from the mid-nineteenth century onwards, although coffee has its best sites on the slopes of the southern *sierra*, up to heights of 2 000 metres. Rice and sugar cane cultivation, too, have played an increasing role. Thus, the cultivation pattern of the

Fig. 6.8. Ecuador: vertical land use profile (after Sick, 1959)

The profile runs roughly southwest to northeast through the coastal foreland and the high mountains, and it indicates the sequence of the distinct economic regions with their main products. The relationship of these with altitude, temperature conditions and annual distribution of rainfall is obvious. The lowest zone is the *tierra caliente* (up to 1 000 m): just inland of the Pacific coast, which is covered by mangroves or coconut palms, is the thorn savanna, extensively utilised for stock raising. Above it rise the 500 to 600 metre high *cerros*, which are more wooded since they receive rainy mists (*garuas*) even during the dry season. On the fertile alluvial river embankments (*bancos*) grow the country's important export products (coffee, cocoa, bananas). On the wetter part of the river margin (*vega*) and in the depression between the *bancos* (*tembladeras*) rice is grown. Besides rice the *tembladeras* are used for stock raising during the dry season; during the rainy season the stock find pasture on the surrounding hills (*lomas*), out of reach of flood waters.

Above the *tierra caliente*, at an altitude of 1 000 to 2 000 metres and with a mean temperature of 18 to 22°C, there is the *tierra templada*, which comprises both the transverse Andean valleys (growing coffee, sugar cane and bananas) and the humid mountain forests of the Pacific slope, which climb up into the *tierra fría* (2 000 to 3 000 metres) as cloud forests.

Up above this is the *tierra helada* (3 500 to 4 700 metres). Here, potato and barley cultivation, and also stock-raising, reach their altitudinal limits between 3 500 and 4 000 metres. The snowclad summits of the High Andes belong to the *tierra glaciale*.

The diagram shows the altitude limits of the most important cultivated plants.

coastal plains and the lower valleys of the high mountains of the *sierra* is very varied.

The division of Ecuador into three parts i.e., the *costa*, the *sierra* and the *Oriente*, stresses regional contrasts which find expression in both the physical geographic and socio-economic environments. The *costa* which comprises both rain forest and dry savanna, has a much higher proportion of negroid population, and the active *montuvios* differ markedly in temperament from the highland Indians. The inhabitants of the *sierra*, who are economically very dependent on the *costa*, have remained isolated in many parts right up to the present, and have remained conservative in their politico-social structure. However, the *sierra* is the traditional core area of the country. Until the recent oil discoveries the *Oriente* region, situated off the beaten track, did not have much economic importance. The forest tribes of this area have been even less incorporated into the national culture than those of the Andean region (Preston, 1965; Bromley 1972).

In the American tropics large numbers of the population were hesitant for a long time about settling in the more low-lying habitats and it is only recently that they have increasingly begun to move to the wooded lowlands. Trends similar to those mentioned above in the case of Ecuador can also be observed in other Latin American states (e.g. Costa Rica, Colombia, Peru and Bolivia). With a rapid population increase (annual growth rates mostly over 3 per cent) there has been great population pressure on many intra-Andean high plateaus, leading to penetration into the lowlands, which to a large extent were still unused. Often the people settle as squatters or homesteaders on private or public lands. In the frontier zone the forests have been pushed well back by uncontrolled clearing. Although in these unplanned types of peasant colonisation, men with billhooks have simply penetrated into the wilderness, usually well in advance of the roadmakers, reclamation often starts from old-established settlements along well laid down routes. In this kind of development, the economic status attained has often been little better than subsistence. In other Central or South American territories extensive areas of virgin land were acquired by the financially strong ruling upper class for very little money, either to settle tenants or colonists, or to set up plantations.

Multiracial societies and the agricultural structure of Malaysia

In monsoon Asia the cultivated areas of the wet summer temperate zone continue into the tropical zone without there being any break in the form of a broad arid zone. Intensively cultivated gardens and skilfully constructed terraces are typical of the way in which the natural landscape has often been radically transformed by agriculture. A dense and racially often heterogeneous rural population, together with well-developed urbanisation, are characteristic of many parts of Southeast Asia.

The heterogeneous structures of society in these countries cannot be explained only in terms of the accidental coexistence of different national, religious and language groups. These groups influenced each other considerably. They often only settled in the later period of colonial history. This is clearly reflected in the agrarian cultural development of Malaysia.

Malaysia falls into three broad economic regions: the more densely populated western coastal lowlands, the deserted mountain rainforests of the centre—still insignificant economically—and the rather inaccessible eastern coastal areas. At the present day, the remains of an old hunting and food-collecting people, as well as peasant groups practising shifting cultivation, have been pushed back into the central rainforest. In the other agricultural regions we find, alongside the dominant Malayan population, other national groups, mainly Chinese, Indian and (in far fewer numbers) European. As far as international markets are concerned, the plantation economy is particularly important, as well as tin-mining (Uhlig, 1963).

European plantation farming started much later here than in the New World, where slave labour had facilitated the practice. It was only towards the end of the nineteenth century that shoots of the *hevea* rubber tree, which is used in the collecting economy of the Brazilian rainforests, were brought to Malaysia, creating the basis for the development of the country's rubber production. Since the Malayans themselves showed little inclination to give up their subsistence economy and hire themselves out as labourers, the rubber plantations were dependent on Chinese labourers and later, to an increasing extent, on immigrants from southern India. The monotonous but clean and hygienic plantation settlements became new homes for these workers, who were brought to this country by the managers of predominantly foreign-financed companies. Even today, almost half of the Tamils, most of whom have been born in the country, work in agriculture. They are concentrated mainly in the plantation states of the Federation (e.g. Selangor). On the other hand, the Tamils play but little part in rice and vegetable cultivation. The proportion of Indian labourers has been decreasing recently, since as living standards improve they are being replaced in this work by the Malayans.

For a long time the agrarian structure of Malaysia was characterised by the predominance of rubber cultivation on large plantations. It is only recently that this predominance has been diminished by the expansion of rubber cultivation on smallholdings. Moreover, the number of plantations in foreign hands is decreasing through sales. Such plantations are usually parcelled out by the state and the smaller properties thus obtained sold to Malayans since, for political reasons, the Malaysian government is not keen on letting them be bought up by rich Chinese. This procedure meets difficulties, however, because of the poor financial circumstances of the Malayans.

Nevertheless, a large proportion of the smaller-scale rubber plantations is in Chinese hands, since parts of subdivided large properties had already been acquired by Chinese. Moreover, many smallholdings which were originally Malayan-owned come into the hands of Chinese merchants when they are resold. More recently, cooperatives have developed out of the process of parcelling out plantations to create small-scale enterprises. Several smallholders, who only own 2 to 4 hectares each, will decide to cultivate a larger plantation area together, in order to be able more profitably and more rationally to employ relatively expensive modern mechanical means of preparing the area to be cultivated.

The Chinese were the first to go in for rubber cultivation on small plantations and, in part, on individual farms (Fryer, Jackson, 1966, pp. 198–228). With British help and instruction the Malayan farmer also followed this example. Chinese farmers often carry out rubber cultivation alongside their other activities as traders, buyers, miners or market-gardeners.

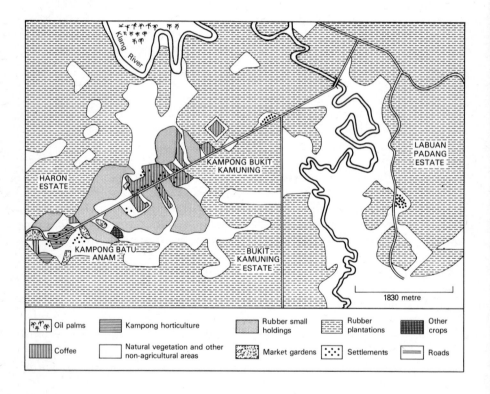

FIG. 6.9. Land use in the lower Klang valley, between Kuala Lumpur and Klang, Malaysia (after Ho, 1962)

In the Klang valley smallholdings are limited either to areas in which rubber and coconut cultivation is diminishing or else to poor soils for which only low taxes have to be paid. The lower Klang valley is swampy, with pools and dumps of spoil bearing witness to former tin-mining.

Compared with these plantation cultures (among which tea and oil palm also play a part and small-scale peasant plantations, the cultivation of food-crops and fruit by smallholders is insignificant. To a large extent it consists of the cultivation of rice, which is carried out mainly by Malayans on farms of 2 to 3 hectares. The Chinese, on the other hand, are particularly successful vegetable growers since, in contrast to the Islamic Malayans and Tamils, they keep pigs and thus have manure at their disposal for their plots.

The most striking feature of Malaysia's agrarian structure is the contrast between the more densely populated and economically active western areas, with their tin-mining carried out by the Chinese and Europeans, and their rubber plantations, and the more thinly populated east, mainly worked by Malayan rice farmers. The coastal strips and the river courses have remained the principal cultivation-areas of the Malayan rice farmers. The foci of these 'old' Malayan agrarian regions are the *kampongs*, either loosely scattered or in long rows, with their farmhouses on stilts above the water and their kitchen gardens in the shade of groves of trees. They lie between the *sawah* (wet rice land, usually near a river) and the swamps and forests.

The monarchic organisation of rule by sultans and rajahs, which originated under Indian influence, was of great importance for the development of Malaysia's agrarian structure. One-tenth of the harvest belonged to these feudal lords. As in other parts of Asia this system became a considerable burden on the peasantry, as middlemen and tax lessees increasingly intervened (Schiller, 1964, p. 83). When property registers were introduced in the second half of the nineteenth century, many ancient claims to land were established. Alongside these, however, traditional forms based on unwritten laws are still important, so that the circumstances of landed property and tenancy, especially in northern Malaysia, are extremely complex. The high proportion of rented land within the total would seem to indicate that land reform is very necessary; indeed, the first step in this field was the rent law (Padi Cultivator Ordinance, 1955).

A rural cooperative system has been successful, mainly among the rice farmers of the Malaysian population. As in other parts of Asia the private moneylender, who in Malaya is usually Chinese and who charges high interest rates, plays an important part.

The balance of demographic strength between over five million Malayans and the nearly equally numerous Chinese is very unstable, quite apart from the Indian and European minorities. At the moment the Malayans still represent marginally more than half the total population; but the state, in its constitution, guarantees them three quarters of all the administrative posts and recruitment into the army.

The island of Penang, which lies off the west coast of Malaysia and which used to be a British trading base is another good example of the strong intermingling of different population elements. At the time of the foundation of the settlement, as a free port, the island was uninhabited. Today it is the dwelling-place of more than 300 000 people. The population of the principal town, George Town, consists of groups of different ethnic origin. Corresponding to a vertical social stratification, which is based mainly on income, there is a spatial side-by-side arrangement of individual national groups, both in the town and in the countryside. In this arrangement the absolute numerical majority is held by the Chinese in the strongly urbanised northeast of the island (George Town and its surroundings); the Chinese have immigrated from various parts of southeast China since the beginning of the nineteenth century. The Malays, on the other hand, only represent over half the population in the rice-growing areas of the southern and western alluvial plains. The Indian element, living in George Town and the rest of the northeast, makes up over 10 per cent of the population and is mainly occupied in the trade and services sectors.

West of George Town the isolated farms of the Chinese market gardeners predominate in the hilly country, whereas the linear and mainly Malay-occupied villages characterise the wet-rice *sawahs* and mixed tree cultures. Chinese and Malay national groups live alongside numerous Tamils in the fishing villages along the north coast of the island (see Fig. 6.10, and Küchler, 1968).

Multiracialism and a population which shows a strong traditional tendency in its religious adherence (e.g. to Buddhism) are also characteristic of the social and economic structures of other countries in southern Asia.[1]

Racial and religious heterogeneity are marked features of many tropical countries, and developed out of events in the history of civilisation and the world economy such as the division of labour between the inhabitants of tropical regions and the inhabitants of the temperate zone industrial countries, or the acquisition of suitable labour forces during the colonial period. Heterogeneity of social structure is not limited to former colonies but can also be observed in countries which, like Thailand, were never colonies. What is particularly striking, on the other hand, is the high degree of assimilation of multiracial societies on tropical islands (such as Hawaii,

---

[1] In other parts of Asia the differences in the pattern of agricultural development are attributable to adherence to distinct religious congregations. Wirth (1965a/b) illustrated the sociogeographic conditions among different religious communities in the East. He quotes as an example the Maronites in the Lebanon, whose position in comparison with neighbouring Christian and non-Christian groups is striking.

Many agricultural innovations such as, for instance, the cultivation of table apples and the construction of groundwater wells, were initiated in the Lebanon by the Maronite population. Other innovations such as, for example, irrigated groundnut cultivation on the Syrian coastal strip, go back to the activities of Islamic communities who had a closer connection with Western ideas. In the Europeanisation and Americanisation processes so typical of many developing countries the imitation of modes of conduct carrying higher social prestige is particularly important. The influence of Lebanese and Syrians who had emigrated to the USA, Latin America or West Africa was very important in the economic development of their home countries.

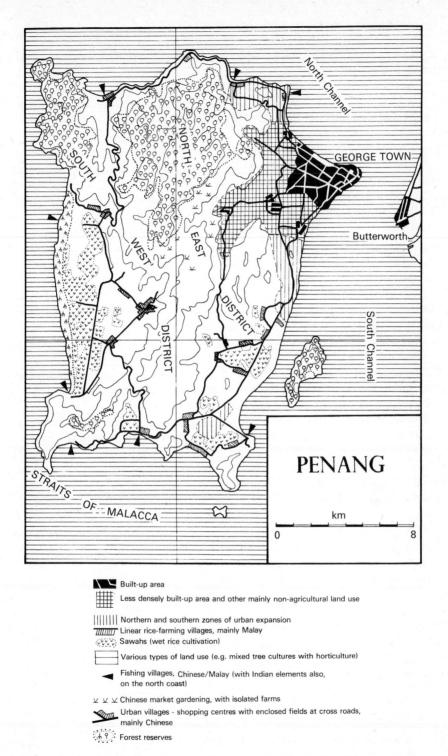

Built-up area

Less densely built-up area and other mainly non-agricultural land use

Northern and southern zones of urban expansion
Linear rice-farming villages, mainly Malay
Sawahs (wet rice cultivation)

Various types of land use (e.g. mixed tree cultures with horticulture)

Fishing villages, Chinese/Malay (with Indian elements also, on the north coast)

∨ ∨ ∨ Chinese market gardening, with isolated farms
Urban villages - shopping centres with enclosed fields at cross roads, mainly Chinese
Forest reserves

FIG. 6.10 Cultural spatial classification of Penang (after Küchler, 1968)

Mauritius and Madagascar) where this feature is also clearly reflected in the agrarian structure.

Tea plantation in Sri Lanka (Ceylon)

The central highlands of Sri Lanka (Ceylon) provide an interesting example of a complete change of landscape following the introduction of tea plants. Within southern Asia there are few regions that demonstrate this striking change so clearly. Only the mountain scenery and macroclimatic conditions—high rainfall values—seem to have remained unchanged. Other components of the existing ecosystems were deeply affected by the turnover to tea planting.

In the history of Ceylon, three main colonial periods may be distinguished: Portuguese (1505–1658), Dutch (1658–1796), and British (1796–1948). All these have left imprints on the island. With Christianity new legal ideas and among other innovations the system of plantation agriculture were introduced by Europeans penetrating the island from the coast. In 1815 the last Singhalese king surrendered and the access into the interior highlands was open. Before the British conquest of Kandy (1815), the central highlands had remained in a state of primeval wilderness. Many changes date from about this time. Early big game hunting was soon followed by commercial interests.

Coffee, which was known through Arabic influences, was successfully introduced. It transformed a considerable part of the highlands up to an altitude of 1 200 m. In the late 1860s and 1870s a fungus devastated the existing coffee plantations within the span of a few years. Experiments with cinchona failed, because of low prices on the world market. (In 1884 about 25 000 ha were planted with cinchona, mainly in the southeastern mountains.)

After this experience the stage was set for the introduction of tea. Although earlier experiments with tea had been carried out in the first half of the nineteenth century, it was only after the introduction of hybrids from Assam that the first larger estates were established. The following figures give an idea of the rapid expansion of tea culture: 1867, 4 ha; 1875, 40 ha; 1885, 40 000 ha; 1905, 160 000 ha; 1960, 240 000 ha. These figures clearly show that coffee was not only replaced by tea, but larger and hitherto uncultivated areas were taken over. While the upper limit of coffee cultivation was around 1 200 m, tea could be planted up to over 2 100 m. It is being planted today on three main levels: up to 500–600 m as a local cash crop in peasant cultivation; from 1 000–1 200 m in a transitional zone of a dual economy between peasant agriculture and plantations, with ricefields in the valleys and tea on the slopes; above this altitude tea is grown in monoculture. Here the whole economy depends on it.

In the neat tea gardens the shrubs are being cut to a compact size of about 1 m, so that they can be plucked easily by the Tamil women. Terracing and drainage ditches prevent serious soil erosion. In the factory buildings the rather complicated processing of the green tea leaves takes place. The labour lines of the plantation workers and the bungalows of the administration are typical for the settlement pattern of this area. It is remarkable that all these components of the regional structure were originally alien to the country.

This is also true of the people introduced as coolie labour by the British. There are two major groups of Tamils in Sri Lanka numbering over 1 million each. The Indian Tamils work mainly on the plantations of the Central Highland; of lowest caste and as Christians, they have come from South India only within the last century. The Jaffna Tamils of higher caste are often Hindu. They are descendants of the first Tamil immigrants in the north. There exists considerable political unrest between the Singhalese and Tamil population. After independence legislation (e.g. about language and administration) was often directed against the Tamils. Thus the intake for estate labour from outside Ceylon during British rule has created population problems which are of profound importance to the island and which also affect the backbone of the island's economy: the plantation agriculture of tea (Schweinfurth, 1966, 1971).

Marby (1971) has studied the ecology of the southwestern low country and of the central highlands of Ceylon, using the medium of the tea plant. The fact that the same plant is growing at all altitudes between sea-level and 2 250 m under the obviously different conditions was taken advantage of in studying the *Teelandschaft* in order to gain better understanding of the country's ecology. Interest in his study was greatly enhanced by the fact that the tea plant was originally a stranger to the island and so everything connected with its cultivation had to be introduced. Brought in as a plantation crop after the decline of coffee cultivation and with a real demand for it on the world market, tea spread rapidly to fill the 'vacancies' left from the coffee days and turned hitherto unspoilt montane rain forest and uninhabited hill country into a manmade environment. In the end it established itself as the main support of the island's economy. The initial success of tea led to its further extension, by which time it had evidently found a congenial habitat. All land used for tea cultivation was first cleared of natural or secondary forest unless it had already been cleared for coffee estates. The natural vegetation of more than 200 000 ha was destroyed and a highly controlled foreign plantation system took its place.

Tea did not remain the only foreign element introduced. It was followed by various others so that the tea-growing areas are entirely made up of elements originally foreign to the island: shade trees, more often than not were of Australian origin; garden plants and trees, bushes, flowers, and vegetables from all regions of the former British Empire, but mostly European, Australian or South African. Besides the entirely new imported vegetation, a new human element was introduced. With few exceptions, owners and managers were British, and the labour force came from South India.

Tea and all the other imported elements had, somehow, to fit in with the existing ecology; but they greatly affected the chosen sites and areas as well. The tremendous altitudinal range offered by this tropical island within short distances by virtue of the existence of the central highlands explains why such an astonishing variety in 'challenge and response' can be seen.

The originally foreign, imported form of land use has become a part of the island, and today the island's economy depends on it. But the tea cultivation itself is in turn, as much influenced by Sri Lanka's economy and politics as it is by its natural environment (Marby, 1971).

A comparison of two economic regions: Burma and Thailand

Hettner once called regional comparison the 'geographical experiment'. This comparative method has been used in numerous studies (e.g. Krebs, 1951). In the tropics, too, interesting areal parallels and differences can be shown; bear in mind, for instance, the differences in the development of the Congo and Amazon basins, the Niger and Ganges deltas, or the Namibian and Atacama deserts. A comparison of two neighbouring countries in the Asiatic tropics—Burma and Thailand (Siam)—is instructive for assessing the agrarian situation in tropical countries.

A glance at the map (Fig. 6.11) reveals certain similarities of position and physical structure between the two countries. In addition, both Burma and Thailand are relatively thinly populated in comparison with their two densely populated neighbours India and China. This is mainly due to the strong isolation imposed by the high mountains running north to south along the border. The frontier with India is formed by the West Burmese mountain range; Thailand is screened from Burma by the Central Cordillera; the Mekong marks Thailand's border to the east. The Cordilleras along the border decisively delimit the structure of the two countries. They keep away the monsoon rains from the central regions and thus cause the formation of large, relatively dry inland zones in their rainshadow. They also account for the long north–south courses of the rivers which, on the other hand, reflect in their discharges the level nature of their catchment areas. These 'vital arteries' are the Irrawaddy and Salween in Burma, the Menam and, on the eastern border, the Mekong in Thailand. Both in the dry inland areas and at their mouths these rivers have deposited fertile alluvial plains.

The high frontier mountains not only protect the two countries from each other and from their neighbours India and China, they also form a barrier cutting off the two central alluvial plains. Both the Shan Highlands, for instance, on the eastern frontier of Burma, which have a mean altitude

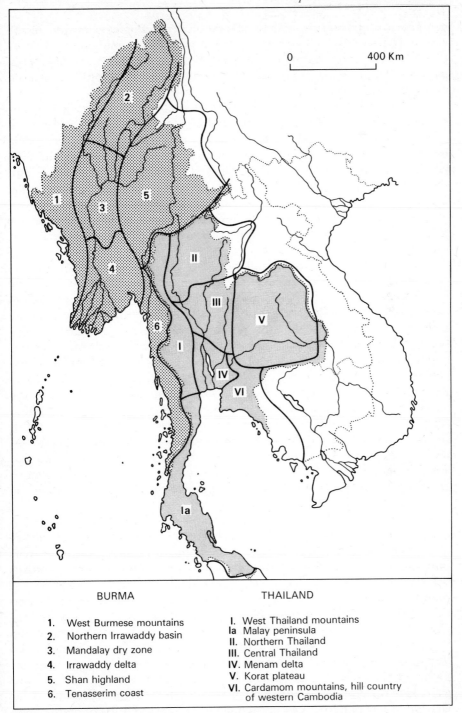

0   400 Km

BURMA	THAILAND
1. West Burmese mountains	I. West Thailand mountains
2. Northern Irrawaddy basin	Ia Malay peninsula
3. Mandalay dry zone	II. Northern Thailand
4. Irrawaddy delta	III. Central Thailand
5. Shan highland	IV. Menam delta
6. Tenasserim coast	V. Korat plateau
	VI. Cardamom mountains, hill country of western Cambodia

FIG. 6.11. Physical regions of Burma and Thailand (after Maas, 1963)

of 1 500 to 2 000 metres, and the Korat Plateau (200–300 metres) in the east of Thailand, which consists of fine granular mesozoic sandstones, are cut off from the alluvial plains by sharp precipices which have hampered the opening up of the highlands and retarded these regions' integration into their respective nations.

Parallels are also obvious in the vegetation covers of the frontier mountains of both countries: in the north and northwest, in each case, they are covered with deciduous monsoon forests in which teak is of special economic importance. On the plateaus of eastern Burma mountain forests dominate, whereas the vegetation of eastern Thailand is similar to that of the humid savannas, because of the lower altitude. Both countries have a share in the Malay peninsula, which is of considerable economic importance to them because of its rich tin and wolfram deposits. It is also a favourable area for rubber plantations as it lies within the sphere of the perpetually humid tropical climate.

The distribution of the population and the spatial structure of the economy depend largely on the physical features. The spatial pattern of the economy also reflects the structural relationship of both countries. Moreover, the corresponding subregions have similar functions within the national economy. The old cultural core areas are not, as might be supposed from present conditions, the several delta areas but the adjoining dry regions to the north: in the case of Thailand the region around Chiangmai and Sukothai (Fig. 6.12), in Burma the area around Mandalay and Prome. Relics of old irrigation systems and historic capitals are witnesses to the former importance of both these regions. Following the river valleys, Burmese and Thais migrated into the arid regions under Chinese pressure in the first millennium AD. In the relatively dry areas they found good conditions for irrigated rice cultivation. In the last two centuries, however, these economic regions have lost importance within their respective countries, although many products (legumes, cotton, rape seed) can be grown and rural life had developed well there.

The chief economic centres today lie in the delta regions to the south. In Thailand this shift from the drier areas began with a continual expansion of the cultivated areas towards the south. Population increase was a great impulse to this movement. Moreover, the development of the Menam delta was instigated by the strong demand for rice on the international market from the middle of the nineteenth century onwards. In Burma, on the other hand, it was a more exogenous impulse—that is to say, the inclusion of Burma into the British Empire after 1855—which caused the colonisation of the Irrawaddy basin. Especially after the opening of the Suez Canal in 1869 Great Britain was interested in opening up new markets for her textiles in Southeast Asia. The growing of rice as an export crop was fostered at the same time, and the rice-growing areas in the Irrawaddy delta were greatly expanded. A subject which has remained quite obscure is how the swamps were reclaimed; the marshes are very difficult to cultivate and

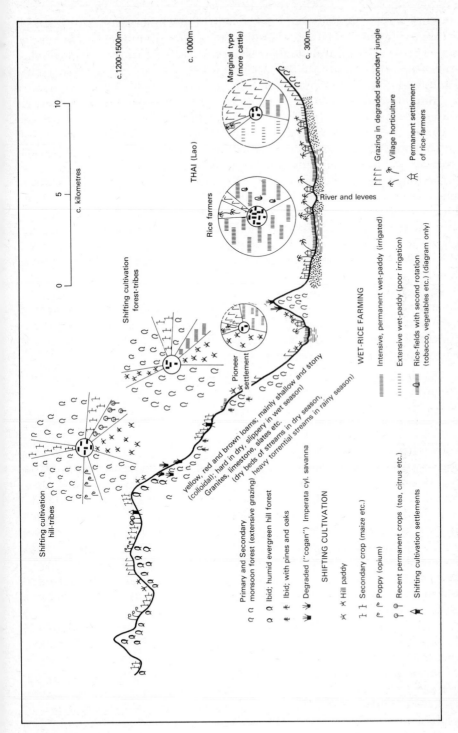

c.1200–1500m.

c. 1000m.

c. 300m.

Marginal type
(more cattle)

THAI (Lao)

Rice farmers

c. kilometres

Shifting cultivation
forest-tribes

Pioneer
settlement

River and levees

Shifting cultivation
hill-tribes

yellow, red and brown loams; mainly shallow and stony
(colloidal); hard in dry, slippery in wet season)
Granites, limestone, slates etc.
(dry beds of streams in dry season,
heavy torrential streams in rainy season)

Shifting cultivation
hill-tribes

⌒⌒ Primary and Secondary
monsoon forest (extensive grazing)

ᠹᠹ Ibid; humid evergreen hill forest

⚹⚹ Ibid; with pines and oaks

⚹⚹ Degraded ("cogan") Imperata cyl. savanna

SHIFTING CULTIVATION

⚹⚹ Hill paddy

⌐⌐ Secondary crop (maize etc.)

ᠹᠹ Poppy (opium)

ᠹᠹ Recent permanent crops (tea, citrus etc.)

◀ Shifting cultivation settlements

WET-RICE FARMING

▨ Intensive, permanent wet-paddy (irrigated)

▨ Extensive wet-paddy (poor irrigation)

⊞ Rice-fields with second rotation
(tobacco, vegetables etc.) (diagram only)

ᠹᠹᠹ Grazing in degraded secondary jungle

ᠹᠹ Village horticulture

⛪ Permanent settlement
of rice-farmers

Fig. 6.12. Schematic landscape profile and land use diagrams: Northern Thailand (after H. Uhlig 1969)

the work was carried out without the help of machinery, succeeding solely thanks to the manual labour of millions of people. Simultaneously with the drainage of the swampy delta there was a considerable movement of population from north to south. In the 1880s there was actually talk of a 'rice rush'. The population of the south rose from 2·6 million inhabitants in 1872 to 5·4 million in 1891. In 1911 it reached 6·2 million and today more than 11 million people live there (the total population of Burma in 1970 was 28 million). The rapid development of the Irrawaddy delta led to important agrarian and social regroupings. The change from a subsistence to a market economy was bound to bring about a transformation.

Economic setbacks caused many rice farmers who had previously been independent to incur debts. Gradually a tenancy system developed. As in Thailand, these tenants could obtain leases of only short duration. It is only in the last two decades that the Burmese government has carried out agrarian reforms. In some planning projects collective farms were created in various rice-growing areas which are worked jointly by five to twenty families.

The new main economic regions in the delta and coast areas, including the Arakan coast, are strongly orientated towards world markets because of their favourable location on traffic routes. This is also reflected in the rapid growth of the new capitals Bangkok and Rangoon compared with the old centres such as Mandalay, for example:

1892 Rangoon 170 000 inhabitants, Mandalay 176 000.
1931 Rangoon 400000, Mandalay 134 000.
1971 Rangoon 1 927 000 (with suburbs), Mandalay 360 000.

In particular, Rangoon's development into one of the world's main ports for rice exports was closely connected with the reclamation of the Irrawaddy delta. It is typical that the major part of the population of these ports is not represented by the native population but by a colonial middle class. In Burma these were mainly Indians, among whom the *chettyars* ('lessors' or 'moneylenders') held a leading position as rice dealers. A large section of the Indian population returned from Burma to India after the Second World War. In Thailand the same function was carried out by Chinese, who right up to the present hold important positions as controllers and middlemen in the economy of the country, including the plantation system and mining.

The third important economic region of the two countries is situated in the south, on the Malay Peninsula: Tenasserim and southern Thailand. Ever since prehistoric times tin-mining has been important here. The state of industrial development in both countries was so low that previously the ores had to be exported to Malaysia; now, however, there is a modern tin smelter at Phuket. Since the transport system has been well developed in response to mining activities, the creation of numerous rubber plantations has been facilitated. After the deltas, these peninsular areas are the best-developed economic regions of both countries.

The 'active areas' described so far contrast with really 'passive areas' in the north and east, and west and northwest. The lowest population densities are found in the northern and northwestern mountains, which are but thinly occupied by various tribes in conditions of retreat. They practise an extensive land rotation economy. The combination of slash-and-burn agriculture and destructive exploitation by private forestry companies has been responsible for the marked decline of stands of teak in the monsoon forests. A variant of the land rotation economy, the *taungya* system (see pp. 61 and 110), was introduced here, but with only partial success.

The two regions in the east, the Shan Highlands and the Korat Plateau, have been sources of worry to both governments. Their economic development has been hampered to a large extent by unfavourable transport conditions and tribal quarrels. So far, subsistence farming has predominated, with extensive stock-raising and mountain rice cultivation here and there.

Both Burma and Thailand are thus agriculturally orientated countries, which up to now have depended on the export of a few raw materials (rice, rubber, tin). The civilisations and cultures of both countries were fundamentally influenced by Buddhism.

Two further sociological phenomena hamper steady economic growth: the former colonial middle classes in the towns enjoy an unduly large economic importance in relation to their share in total population; secondly, the large number of minority groups living in the marginal, thinly-utilised mountain regions, have hardly been incorporated into the economic system. In comparison with their neighbours India and Indonesia, however, Thailand and Burma have managed to achieve a modest prosperity for their people, even though in both cases the economic structure is almost entirely agricultural. Agricultural production can still be increased by enlarging the cultivable area through sensibly applied irrigation programmes and intensive manuring.

The two countries resemble each other both in their economies and in their agricultural spatial patterns. This is due to natural endowments and demographic development, which is closely linked to the north–south migrations of different ethnic groups and to the effects of alien intermediate classes.

Problems of land tenure in tropical Africa[1]

In the study of the social and economic patterns in the tropics, not only does the question of land utilisation arise—which has already often been studied by geographers—but also the main issues of agrarian structure and land

[1] See Manshard, 1965, a, b.

tenure[1] This applies to the tensions between *latifundios* (large properties) and *minifundios* in Latin America and to the rent-capitalist structure of the Orient, even more than to tropical Africa.

Geographers can contribute to the study of land tenure conditions because it is one of their main concerns to relate regional differences in natural environmental conditions and socioeconomic structures; it is precisely the ownership of land which places man in a definite, spatially fixed field of activity and function.

The agrarian codes of most African peoples are extremely varied. Ownership, tenure, settlement and land utilisation rights cannot easily be explained according to western ideas, because of the way in which economic, social, religious and political factors are interwoven. Ethnologists and social anthropologists have made reference to the close links between the land and human society, which in the broadest sense includes ancestors and descendants as well as contemporary family units.

This relationship is expressed in numerous formulae: for instance, as *maître du sol* or *chef de terre* (Biebuyck, 1963; p. 52).

The rights referred to above may be vested in the village community in its totality; the extended family; the clan (usually represented by its elders); the individual (especially in more recent times); or even, as in the old days, in the chiefs, the feudal lords or the priests (for instance, the 'earth priests' or witch doctors).

Common forms of rights to land in African agrarian law are: permanent possession vested in the rightful owners of property; land tenure with a time limit; different types of rights of use and partial use. The right of partial use is exercised in southern Nigeria, for example, in such a way that the lessee retains the right to make use of fruit trees while conveying to the tenant the right to use the soil.

The first signs of land speculation commonly only appear in systems with individual ownership, whereas with communal property sales or mortgaging are not usually possible because of the magical religious principle of the land being inalienable. In its spatial expansion the individualisation of landed property often followed on European influence, spreading its effects from the coastal areas into the hinterland.

In this context, land tenure conditions are good indicators of population increase or decrease, and of the superimposition of various groups belonging to different economic stages and systems. Often there are varying ideas or concepts of land tenure in similar agricultural regions. Moreover, modern economic thinking has exercised a great influence. Modern, rational and older traditional influences have created the variety of agricultural systems and agrarian customs to be seen within tropical Africa; this is

---

[1] Jomo Kenyatta (1938, p. 21): 'Land tenure is the most important factor in the social, political and economic life of the tribe.' For the definition and delimitation of the terms 'agrarian structure', 'land tenure' and 'agrarian constitution' see Ringer, 1963, p. 13.

reflected nowadays in the variegated colourful scene of agricultural land-scapes, ranging from seminomadic shifting cultivation to irrigation and the European plantation economy.

These differences in degree of economic complexity, ranging from sub-sistence to a world-market orientated agriculture (cash crop cultivation), have had profound effects on all land tenure structures and land use methods. The time factor is of considerable importance. At one time there was generally ample land available for everyone, and it was controlled by the group rather than the individual. Thus, at first, apart from its import-ance in providing subsistence, the land did not have any special economic value in the sense of its being exchangeable or saleable. Nevertheless, land ownership implied prestige. Quarrels between groups about land, which are so typical everywhere nowadays, were formerly far less common. Naturally, disputes between tribes frequently occurred in the past, in the course of African population migrations, when the conquest of land was very important.

Ringer (1963, p. 179 ff.) characterised the basic principles of African agrarian constitutions, which so far have not been codified, in the following concise terms: right of first occupance; the concept of control by one group; right arising out of usage. Outsiders who are not group members can be admitted so long as they observe certain duties in the use of the land. Here one has to take into account the value concepts of the groups active in agriculture, concerning the cultivation of certain crops. There are special rules for the use of trees and bushes, fruit orchards and palm groves, be-cause even after the African farmers have handed over the areas they have cleared for the annual crop cultivation they retain the right to collect the products of the trees.

The distribution of land was, and still is, in part, the task of the king, the chief or the head of the family. This principle was even observed in the assignment of grazing rights among the nomadic tribes who practised com-munal ownership of livestock. It is a principle ranging, therefore, from farming groups to pastoral nomads. The closely interwoven basic prin-ciples of these agrarian customs have to be borne in mind when attempting to institute land reforms. A crucial question for economic development in the African continent is the difficulty of making a regional delimitation of widely differing and complicated agrarian customs, and distinguishing one from another. Although there exists an extensive and mainly ethnographic literature on these subjects, few works have studied African agrarian cus-toms thoroughly. In order to be able to analyse exactly the present situation and the future prospects for development of African agriculture, the geographer has to try to investigate the intricate and variegated structures of different agrarian systems. To make matters more complicated, most of these customs are not written down. In addition, Africans often interpret terms such as 'land', 'property' and so on, in a very different way from Europeans.

The topographical representation of the geosphere has been based on rules and methods of mathematics and astronomy which originated in Europe and have found their expression in maps ranging from land register sheets to thematic maps of the earth. Cartography, which was mainly developed by European seafaring nations, bases its projections on certain fixed points and lines. A similar process was applied in the distribution of land in European colonial territories such as North America, with its gridiron pattern of townships and fields, and in South America, with its towns laid out in *cuadras*. In African colonial territories, too, European methods of mapping were introduced in establishing landed properties, despite the fact that these methods were often not suitable for delimiting African landholdings. The boundaries of the latter were frequently altered; they were subject to subdivision; and often even the concept of landed property was quite different (Bohannan 1963).

The Kikuyu in Kenya may be cited as an example. Besides properties owned by individuals or by unions of several clans, there used to exist *rugongos* (mountain ridges). These were political units of property which, in precolonial times, were administered by a 'council of nine', in its turn chosen by similar 'committees of nine' from the next smaller regional organisations (the *Mwaki* or fire units), and by the inhabitants of single homesteads (Leakey, 1952). This indigenous, almost democratic, system was later replaced by the English by a chieftainship system. The 'estates' of such and such a subclan were demarcated and delimited by trees, watersheds, streams, rocks, and so on. Nowadays land can also be sold, provided the other members of the clan are given first refusal. In this way a structure of individual and quite clearly bounded parcels develops which has no longer anything to do with political units.

Especially in East Africa, agrarian codes influenced by Europeans have been superimposed on, or have dispossessed, the old traditional customs. In some cases, as, for example, with the Kikuyu, this was relatively easy. Clearly marked parcels could also be surveyed by photogrammetry and laid down in the property register. In many other cultures, on the other hand, individual and permanently fixed parcels of land first of all have to replace the old units, whose location often changes. These reorganisations can be furthered by the introduction of new cultivation methods.

The consequences which can arise when European methods of delimiting and registering land have been introduced suddenly and quite early into Africa can clearly be seen in Buganda, a province of Uganda (Mukwaya, 1953). Before 1900 the Kabaka (king) was the highest authority in the kingdom on all questions concerning land. The land was administered according to a system which showed similarities to the feudal structure of Europe. These old conditions were replaced by the Uganda Land Settlement agreement of 1900, as well as by the land law of 1908; the agreement was drawn up by Sir Harry Johnston. Traditional conditions were completely ignored or, which was even more likely, Sir Harry Johnston (who had only been in the country a few months) did not know of them. With a stroke of the pen completely new and hitherto unknown property rights were created, which at first could not even be understood by most Africans. A gradual transition from collective to more individual property ownership, such as was more typical in other regions of tropical Africa, did not take place here.

At first, as a result of the Land Settlement, over 20 000 square kilometres of land were allocated to a few thousand chiefs and private landowners. As the exact number of chiefs was not known, nor their status closely defined (and quite a few subchiefs should be taken into account as well), there were only 3 650 names on the first land distribution list.[1] The outcome of several pieces of legislation was the *mailo* system—a corrupt form of the English word 'mile', so called because the allocations were made in square miles. The difficulties of introducing such a system, which demanded an immediate and exact cartographic registration of property, were very great. The exact area of the whole country was not known; even roads did not exist, since at that time Buganda was far off the beaten track. Moreover sufficient technical personnel was lacking. In brief, it was impossible to provide a complete survey of landed properties, especially when they were usually far distant from each other.

The great difficulties which arose in carrying out the cartographic survey of landed properties and settlements were a great stumbling block to the solution of legal, political and administrative problems. These days, the simplest solution is a photogrammetric survey such as was successfully carried out in the Kikuyu areas of Kenya. In Buganda, however, despite the fact that properties came into the hands of individuals at an early date, there are still hardly any demarcation lines (such as hedges and terraces) visible from the air, so that there are only very limited possibilities of surveying by aerial photography. Since the new proprietors could not wait until their land had been measured and registered, there took place everywhere transactions in 'paper acres'—purchases of parcels of land which the land register did not yet contain. These contracts (*engadano*), which only existed in writing, often caused great confusion (West, 1963). In spite of all these difficulties the *mailo* system, which had been imposed on Buganda, proved very useful in the cultivation of coffee, cotton, bananas and tea. This first pioneering accomplishment by Sir Harry Johnston certainly contributed to Uganda's economic stability, although it lacked preparation and faced difficulties at the outset.[2]

[1] In the Uganda Agreement the then regents of the Kabaka received 100–150 sq km each, 20 principal chiefs got over 50 sq km, a further 150 chiefs 20–30 sq km and the majority of owners 5·2 sq km (Richards, 1954).

Generally the history of land tenure systems in Africa shows clearly how much depended on a strong, centrally directed development policy (see also the examples mentioned from East African states such as Uganda and Kenya). This is by no means to say that coercive measures are always desirable; as recent events in tropical Africa have shown, these may bring chaos.

An additional factor making for the success of the system was the availability of cheap migratory labour from neighbouring densely populated Rwanda-Burundi (Richards, 1954), which could be hired by the owners of large estates. If the land had been distributed in parcels of 3 to 4 hectares, as in Kenya today, development would no doubt have taken a different direction. It is interesting to see that in Buganda, too, a certain breaking-up of the original very big estates has taken place. In two of the counties studied by Mukwaya (1953), the number of proprietors increased from 135 in 1920 to 687 in 1950. The average size of properties has decreased from 152 to 30 hectares over the same period. Larger properties still existed in 1950: three-quarters of the country still belonged to the 13 per cent of proprietors who owned medium-sized estates of about 170 hectares (Richards, 1963, p. 274 ff.). If one bears in mind that the area of land worked by one family in many parts of tropical Africa is about 1 hectare, the size of such properties appears considerable.

PLATE 14. Sagitwa (about 3 km south of Kisoro, Kigezi, Uganda). Volcanic cone with adjacent densely populated farmland under intensive use. Banana groves are concentrated at the foot of the volcano and near the compounds. The terraced slopes of the cone are mainly planted with sweet potatoes. The slopes within the crater are very steep, so that the terraces have been abandoned. However, the bottom of the crater is cultivated (Photo: March 1958, flight altitude 3 800 m).

In other parts of East Africa the fragmentation of landholdings presents important problems. The study of the agricultural structure of Kigezi (southwest Uganda) affords an idea of the capacity of an East African mountain region and of the dynamics of an area which for decades (like the southern part of neighbouring Rwanda-Burundi) represented a labour reservoir for Uganda. A study was carried out in the densely populated Bufumbira country in the far southwest of Kigezi. This study will be used here as an example (Manshard, 1965b).

● Round huts	□ Square huts	O Silos or stables for small animals	木 Fruit orchards and shade trees

Fig. 6.13. Property parcelisation pattern in Bufumbira, southwestern Uganda Boundaries of plots (same shading indicates same proprietor) (Sketch: W. Manshard, 1965b)

In the first place, attempts had to be made carefully to map the actual extent of property fragmentation. The physical geographical evidence, especially the greater fertility of the volcanic soils, which have been studied sufficiently in this area and which have often been quoted in the literature, is by no means enough to explain the phenomenon.

With the help of an enlarged aerial photograph (scale 1:4500) the individual parcels in a specific section of the landscape were identified and delimited according to ownership. Bufumbira is an extremely intensively used agricultural region, with a dense mosaic of land use such as is generally only known in the horticultural areas of Southeast Asia. A detailed analysis of the section of map chosen yielded the following results: the roughly 1 000 parcels of the *mutungole* (subdistrict) of Nyarusisa belonged to 412 different people. The same group of people owned a further 1 250 plots within the district, but not marked on the map, plus a further 2 000 fragments of land outside the district (the figures quoted for the area outside the *mutungole* are estimates). The number of parcels within the district was 2 300, so that each landowner had five parcels within and five parcels outside the *mutungole* investigated, on average.

How did this incredible fragmentation of property come about? Besides the high surplus of births and the still prevailing polygamous family structure, it is the inheritance customs (depicted in Fig. 6.14) which are mainly responsible for this atomisation of property in Kigezi. In a monogamous marriage it is the custom for the father to give land to his sons when they get married—while still keeping land for himself and his wife. If the wife survives her husband she can continue to live on his land, but the ownership of the property passes to the nearest heir (usually the brother of the deceased), who now has to look after the widow. If the widow returns to her own family or if she remarries outside her dead husband's clan, the land is shared out among the sons.

In Bufumbira this fragmentation process is accelerated by the custom of sons receiving a plot of land on which to grow marketable crops not only—as in Kigezi—on marriage but also at the time when they become fully-grown workers. It is true that the father retains his right of ownership over such land. After marriage, however, the son receives his own parcel on which to grow cash crops, as well as other fields (*shambas*), often lying at considerable distances from each other, for the maintenance of his family. In polygamous families this property fragmentation goes even further, since each single wife has land allotted to her. In these cases the first wife gets a larger plot than the second, third or fourth wives—who, nevertheless, receive land from the first wife in their turn (Fig. 6.14(*c*)). The husband usually retains the supervision of the whole property. All the women can do is try to obtain a surplus from their land, in order to benefit their children. As soon as the sons of the various wives grow up and marry, they too obtain small parcels of land, which are taken out of their mothers' share. The fragmentation process is thus considerably accelerated and

(*a*) Monogamous family structure (after Byagagaire and Lawrance, 1957) (1): Original undivided property. Part of the land is given to grown-up sons (A, B, C, D) on marriage (2–5), so that the father in his old age (5) owns only a small area of land which, at his death (6) is either taken over by his widow or shared among the sons.

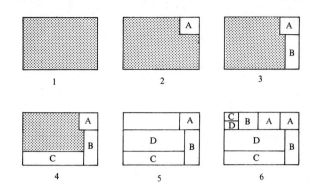

(*b*) Bufumbira system (monogamous): In Bufumbira the sons (A, B, C, D) receive land from their father as soon as they are capable of working properly (*Kwiharika*) (1, 2, A, B, C). It only becomes their own property however, when they marry; at this time (2, A) they also obtain an additional parcel for the maintenance of their family. At the death of the father (3) his property is shared out among the sons (4).

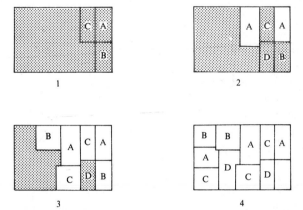

(*c*) Polygamous family structure: In the polygamous system, fragmentation of property takes place even faster. If a man marries a 2nd, 3rd and 4th wife, his property is further and further subdivided. Ever smaller fragments of land have to be divided for the married sons of these marriages (5, A, B or 6, A, B, C, D, E). Difficult social and economic conditions are thus created very quickly (see also Manshard, 1965a).

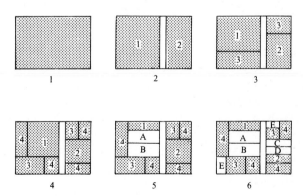

Fig. 6.14. Schematic representation of the fragmentation of landed properties in Kigezi as the result of inheritance systems.

complicated (Byagagaire and Lawrance, 1957; p. 17 ff). One of the most urgent tasks in Kigezi is to slow down and halt the extraordinary property fragmentation. This 'atomisation' naturally has to stop at a certain size of holding, beyond which it cannot continue; at this point, an increased migration into the towns and to central Uganda begins. The same applies to other parts of tropical Africa (e.g. the Cameroons and Malawi). However, such an obvious fragmentation of property is not the general rule in African agriculture. A solution can only be provided through an intensified consolidation of properties.

This widespread fragmentation of the cultivated area into farms of completely uneconomic size calls for reform measures. Certainly, agrarian reforms are an important instrument of economic development; but even when they are thoroughly prepared, they may have rather unpleasant political effects, triggering off a kind of chain reaction.

Again, to bring about a consolidation of landholdings in tropical Africa, by uniting or relocating parcels of land, a number of educational, legal and technical measures are necessary which at this point can only be hinted at, once more using the example of Kigezi in southwest Uganda. With a farming population which is politically at loggerheads and, moreover, as extremely conservative as it is in Kigezi, a unanimous agreement to such reform plans is entirely out of the question. If it were possible, with the help of political parties, churches and other organisations and institutions, to win at least part of the population over to these plans by suitable propaganda, an important first step would be achieved. In such work, visits by politicians and chiefs to other East African areas (such as Kenya) where consolidation schemes have been successful, could be helpful (see below). A further step forward would be the inscribing of the sizes of properties in a land register, especially in the case of large estates; such a measure could have far-reaching consequences for economic development, for example by facilitating loans from banks and savings institutions. In many larger African towns there are already large-scale cadastral maps and exact registers of estates.

After thorough preparations, in which the evaluation of aerial photographs plays an important part in saving time and money, areas suitable for consolidation should first of all be defined and delimited. After deciding on land needed for public amenities such as roads, schools and hospitals (again with the help of aerial photographs), the size of the individual parcels which are to be created should be worked out roughly; then the new distribution should take place. In this distribution, different qualities of soil have to be borne in mind; and suitable demarcation lines, such as hedges or fences, should be found for the consolidated property. All these measures require systematic, scientific and technical preparation, without which only the opposite of what was desired will be obtained. As long as such a consolidation is not possible voluntary mergers have to suffice. The root of the trouble will not be reached in this way. Even more effective would be a change in inheritance customs, which might come with education or new legal measures. With better education, cooperatively organised agriculture might be successful. Measures for consolidating landed properties and preparatory education should certainly go hand in hand.

In recent years the Kikuyu area of Kenya (the districts of Kikuyu, Embu and Meru) has provided a particularly good example of formerly fragmented property being successfully consolidated. Land consolidation in Kenya was connected with the political developments associated with the Mau-Mau uprising. This rebellion by some sections of the Kikuyu tribe was caused, in part, by questions of land tenure; for although the Kikuyu, on the whole, had lost less land to white settlers than the Masai or the Nandi, discontent over the purchase of landed estates by Europeans ('stolen lands') had been growing since the 1920s. With a numerical increase in the tribe from 450 000 in 1902 to over one million in 1948, the Kikuyu region belonged to the most densely populated area of East Africa (Pedraza, 1956, Sorrenson, 1963). The fragmentation of property in the Kikuyu reservation had by then reached an extent similar to the conditions in Bufumbira mentioned above. Here too, it was the prevailing inheritance system which was responsible for a fragmentation which made the economic utilisation of the land almost impossible. At the same time, however, there already existed an influential 'middle-class' group of, for instance, chiefs, government employees and teachers, who had succeeded in building up their holdings by purchasing less densely populated land outside the reservations. Whereas these landowning groups achieved wealth under the protection of the Pax Britannica, the impoverished tenant farmers and smallholders on the Kikuyu reserves proper were forced by the high population pressure to leave the reservations and accept labouring jobs on European estates and plantations. The result was migration to the slums of Nairobi and the reduction of many Kikuyu to the level of a proletariat. In contrast to Uganda, for instance, this development has remained typical of Kenya up to the present day.

In 1954, after the outbreak of the Mau-Mau rebellion, decisive measures were taken to change completely the geographical structure of settlement in the Kikuyu areas. Within a year almost the whole Kikuyu population was moved from their dispersed settlements into compact villages which could be better watched and controlled. This compulsory resettlement had profound consequences. The new villages became nuclei for the development of central functions. Kikuyu who did not own land could live in the villages as manual workers, tradesmen or farmhands, and the first signs of industrial development were soon visible. This resettlement was at the same time a kind of initial stimulus to the later large-scale land consolidation programme. With many individual homesteads destroyed in the course of conflict and upheaval, the parcels of land belonging to any single Kikuyu could hardly be located with exactness any longer. For the Kikuyu who by then were living in compact villages a resiting and consolidation of their land was more practical. This was carried out through years of patient work by various commissions.

By 1965 more than 250 000 hectares had been allocated by these agrarian reforms in the 'White Highlands' (Morgan, 1963; Hecklau 1968). This was done within the framework of the international resettlement programme for African peasants, which took several forms, for example the so-called 'High and Low Density Schemes', with varying property sizes.

The taking over by native farmers of large tracts of European land-holdings in the scheduled areas of Kenya 'under the Million-acre Settle-ment Scheme) has already led to considerable changes in the East African cultural landscape (Belshaw, 1963). It will need another generation yet before it can be assessed whether these land consolidation reforms were beneficial or not. Only then will it be seen whether the Kikuyu inheritance customs, with their urge to divide land, will once more lead to an increasing dissolution and fragmentation. Because of political pressures from the members of the younger generation, such a development is rather more than likely.

The result of this reframing of the agrarian code, that is to say the in-dividualisation of peasant property, was an increased production of coffee, milk and other products, increasing the net income of the farmers or, in-deed, making it possible for them to have an income at all. This essential prerequisite for cultural change in the country created at the same time the basis for necessary improvements such as the terracing of slopes, row culti-vation, advice on the need for artificial manuring and plant protection, the development of the health service and the school system. So far the success of these first steps has been rather promising, and even after the 'Africanisa-tion' of various parts of the White Highlands, quite notable improvements in production could be observed.

In many parts of Africa land tenure and land use systems no longer com-ply with the economic and political concepts of the ruling classes. Apart from special types such as the feudal structures of Ethiopia, certain basic prin-ciples and types of land tenure and land use can be observed, with all their regional variants.

In the past, the traditional land use system of shifting cultivation or land rotation maintained a certain ecological balance and also a certain level of soil fertility. With an increasing population, the introduction of modern technical innovations and the ever-greater penetration of the market economy, we are today witnessing the progressive dissolution of these old practices. In some cases adaptation to new economic growth took place without undue disturbance. In others, as a result of sudden political revolu-tion, it led to the expropriation and nationalisation of land and property— as, for instance, with the seizure of Arab property in Zanzibar in 1964.

With the adoption of world-market orientated cash crops there has gradually developed a different and more economically minded relation-ship between the farmer and his land. He realises the economic value of his property, which goes far beyond the former traditional value associated with status. On the one hand there is a danger that land may be treated purely as a factor of production, within the framework of future land reforms, without sufficient attention being paid to social and historical backgrounds. On the other hand, to be fossilised in old traditions must necessarily hamper economic and social development. Deserving of considerable attention are various experiments, ranging from state-run collective farms similar to *sovkhoses*, through community farms on the Israeli model and European-type cooperatives, to individual free enterprises. Types of enterprise which are centrally controlled by an agricultural collective, on the eastern bloc or Chinese pattern, appeal strongly to some governments of African countries. A small number of trained technicians and a mass of illiterate workers can achieve quicker and better results on such farms than among the fragmented smallholdings usual among more conservative peasant societies.

This study of land tenure and land use problems helps the geographer to gain insight into the complex structure of African economic regions. Descriptions of production, trade and communications alone cannot do this; on the contrary, the sociogeographic and social conditions which characterise an area will have to be studied more closely in order, perhaps, to make possible the mapping of regions with similar land tenure structures.

Agricultural developments in Ghana

A study of Ghana's agricultural geographical development and structure will serve as another example. In tropical terms, Ghana is a comparatively well-developed agricultural country in which farming predominates in the humid forest and savanna zones (Manshard, 1961a, Dickson, 1969).

Most of the early phases of development are no longer evident. The reconstruction of older cultural conditions faces great difficulties. Historical records are scarce, apart from oral accounts, which are often unreliable, and the few notes made by early travellers. Archaeological research is still insufficient. As in Nigeria, the forest belt is usually more densely populated in Ghana, nowadays, than the humid savannas which are immediately adjacent. An extensive part of the forest belt has only been settled in the last few centuries. Even if the humid forests were a kind of barrier, initially, they could not have prevented penetration by man, even in early times. Very recent archaeological finds suggest that the settlement of the whole forest belt took place as a more continuous process than had been presumed up till now. The first people to penetrate were small groups of savanna tribes, but since the fifteenth century settlement has taken place on a larger scale. The intruders brought with them their weapons and tools, and also their political institutions and systems of organisation. New peoples and cultures were superimposed on the family groups of the original inhabitants.

The first settlement seems to have been carried out by small units of hunter-peasants, which were sent out by chiefs and later settled as small farmers. Even nowadays, in West Ashanti, there are some villages in which one finds relationships of dependence and loyalty to quite distant chiefdoms. Place-names, too, are helpful in this connection. Small settlement units often became the nuclei of villages and towns which developed later on. There was a similar development in the area transitional to the humid savanna (the Afram plains). Here the hunter-peasants who had been sent from Kwahu or Kumawu settled in small groups and began farming. Settlement and the acquisition of land continued to spread, later on, within the framework of extensive shifting cultivation, which was practised well into the nineteenth century. At first only small fields were cleared in the virgin forest. After soil exhaustion new areas were sought out. This type of economy also explains the very slow progress of migratory movements, which resembled an infiltration into alien tribal areas rather than large-scale migrations of peoples.

We still find these population movements, connected with shifting cultivation, among the Lobi, the Konkomba and other tribes such as the Bassari, in the humid savannas of the 'Middle Belt'. The penetration of the Krobo into Akwapim, however, is connected with the types of enterprise which were developed in the *Huza* system (Manshard, 1961b). Only in the west of the forest belt (for instance, on the lower Tano) does one find more original conditions in the narrow pioneer fringes of the type which are also characteristic of Amazonia or the Congo.

In the forest belt the Akan and other tribes who came from the dry and humid savannas of the north had to give up animal husbandry, which they had in part practised in the Sudan, and replace it by hunting and simple forms of cultivation. Most of the soils which they newly acquired in the forest were untouched. During this first phase the inhabitants of the forest must have lived mainly on wild fruits and the oil of palm trees, compared with the various types of millet and vegetables which were grown in the savanna. Besides subsistence cultivation a trade in spices and cola nuts developed quite early on, under Moorish and European influences.

Although iron tools and weapons already existed on the Gold Coast before the arrival of Europeans, the agricultural pattern has nevertheless been formed to a large extent by the latter. The Portuguese especially exercised a significant influence on the early economic landscape by introducing maize, potatoes, peanuts, manioc, papayas, pineapples, tobacco, tomatoes and other cultivated plants from the New World. They grew many of these new plants in the gardens surrounding their forts to cater for the soldiers and sailors, and from there the plants were gradually introduced further inland. At the time of the slave trade, from the fifteenth to the nineteenth centuries, a larger number of people often had to be supplied with food on the coast. For this reason the Portuguese, Danes, Dutch and

British promoted an expansion of agriculture beyond mere self-sufficiency in the coastal regions. Because of the great requirements of the forts, agriculture developed farther along the coastal strip from Accra to Sekondi than in the remaining coastal regions. In spite of these first efforts at agriculture, the centuries of colonisation up to the beginning of the nineteenth century were mainly characterised by the gold and slave trades.

In the hinterland, however, the slave hunts had a rather negative effect on the continuous development of agriculture, and even after the overseas slave trade had been abolished widespread tribal feuds caused serious and repeated scarcities of food. Locust plagues, epidemics and the Arab slave trade which continued in the north, also contributed to food crises, although

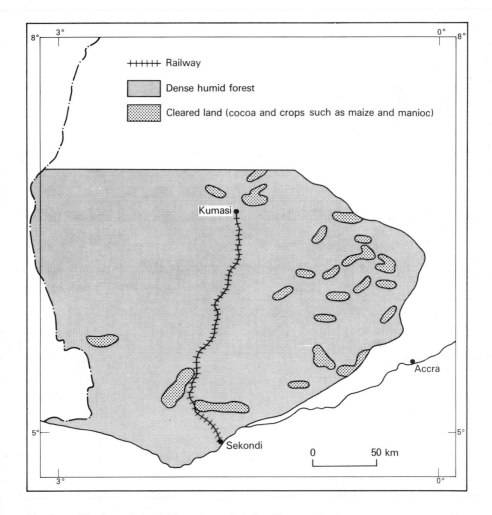

Fig. 6.15. The forest belt of Ghana in 1908 (after Charter 1953)

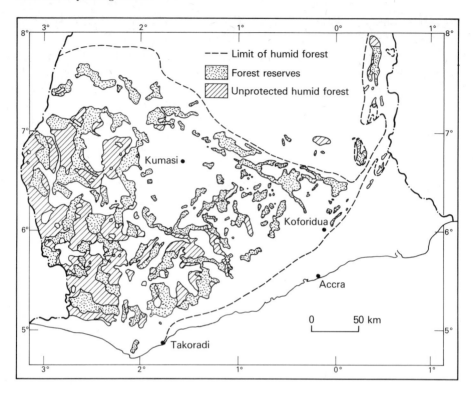

Fig. 6.16. The forest belt of Ghana in 1953 (after Charter, 1953)

these never reached famine proportions. When peace followed the British takeover of the Gold Coast colony, agricultural development proper of the country started, supported by the government and the missions; it brought about exports of palm products, rubber, coffee and particularly cocoa, and it has decisively transformed Ghanaian agriculture since the end of the nineteenth century.

Contrary to the usual views of peasantry, the accumulation of capital in indigenous West African economies has been accompanied by the emergence of specialist farmers. To this class of rural capitalists belong, for instance, the economically motivated migrant cocoa farmers from the *huzas* of southern Ghana. They are really businessmen and not small peasants, who by an unplanned process of trial and error created the world's largest cocoa-growing industry. It is obvious that many stereotyped notions about peasant farmers as a group are quite misleading (Hill, 1970).

With the introduction of these permanent forms of cultivation, and with the extension of the road and railway network and the rapid increase of population (1891, 0·9 million inhabitants; 1948, 4·1 million, 1971, over

8·6 million) there was no longer enough land in Ghana for extensive types of economy such as shifting cultivation. Strict shifting cultivation was replaced by a land rotation economy, with a fallow period between the periods of cultivation. This process is clearly reflected in the landscape. Only twenty or thirty years ago the forest came right up to the cultivated land. Nowadays a wide fringe of open land surrounds all the rural settlements. Areal changes in the landscape are shown in Figs 6.15 and 6.16. It can be assumed that the part of the forest belt which was economically significant, and therefore best known, covered an area of about 80 000 square kilometres at the beginning of the present century. In the more recent past (Charter, 1953, p. 145) forest reserves covered an area of 14 500 sq km, unprotected forests 13 500 sq km, cocoa farms 7 000 sq km, and all the other land used, at least in part, for agriculture (including fallow and unproductive cocoa land) about 45 000 sq km. It can be seen from these figures that only 14 500 sq km of forest reserves and 13 500 sq km of unprotected forest remain of the 80 000 sq km of dense forest cover existing at the beginning of the twentieth century. Moreover, not even this area consists of untouched forest.

In the forest regions of Ghana over 50 000 sq km have been considerably transformed in the last five centuries. Before the introduction of cocoa farming only the land immediately next to the settlements was used for growing plants and fruits for consumption. Cola, funtumia rubber and oil palm were usually only exploited in the marginal areas. As a result of improved communications and an increasing demand for cocoa, the rate of loss of the still unprotected forests rose from year to year, so that there will not be any unprotected forests left in the very near future. As yet, encroachment on forest reserves, which would entail even more serious consequences, has not been considered.

These far-reaching changes within the forest belt have consequences for the climate, soil, vegetation and fauna of which the dimensions can hardly yet be estimated. However, there is no evidence of climatic changes within historic times, since the periods of observation have been too short and reliable climatic data are scarce. Like the majority of tropical soils, the soils of Ghana produce less than those of middle latitudes. The climate causes rapid chemical decomposition, with strong leaching of nutritive materials, and accelerates the formation of laterite and the destruction of humus. It is still uncertain to what extent these conditions can be improved by modern agricultural chemistry and the introduction of suitable fertilisers and cultivation methods. It is much easier to provide evidence of changes in the vegetation. In systematically burning down the forest, the farmers have considerably upset the original balance of nature.

In the mosaic of cultivated areas of Ghana, the cleared 'islands' within the humid forest and the savanna are particularly striking. These cultivated areas also frequently coincide with spatial units of ownership. In contrast to European conditions, the boundaries of cultivated areas are irregular in

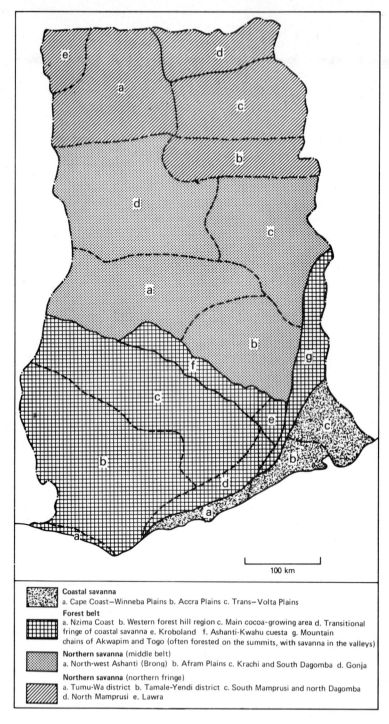

Coastal savanna
a. Cape Coast–Winneba Plains b. Accra Plains c. Trans–Volta Plains
Forest belt
a. Nzima Coast b. Western forest hill region c. Main cocoa-growing area d. Transitional
fringe of coastal savanna e. Kroboland f. Ashanti-Kwahu cuesta g. Mountain
chains of Akwapim and Togo (often forested on the summits, with savanna in the valleys)
Northern savanna (middle belt)
a. North-west Ashanti (Brong) b. Afram Plains c. Krachi and South Dagomba d. Gonja
Northern savanna (northern fringe)
a. Tumu-Wa district b. Tamale-Yendi district c. South Mamprusi and north Dagomba
d. North Mamprusi e. Lawra

FIG. 6.17. Geographical regions of Ghana, with special reference to agriculture
(after Manshard, 1961a, 1962)

hoe cultivation; the latter is generally associated here with quite permanent settlements. Whereas a *flur*-structure, consisting of irregular scattered units, is usually attributed to small *haufendörfer* or hamlets, the occurrence of compact strips of land arranged side by side, belonging to the *huzas* (and companies), is linked with the 'shoestring' village type of settlement (Hill, 1963; Manshard, 1961a, b).

In the northern and southern savannas the total population has increased less than in the forest belt. The richer forest regions, where mining and the timber industry developed alongside cocoa farming, exercised a great attraction, so that the savannas have even registered a negative movement in some parts. In several areas abandoned settlements can be found; the people have migrated to the neighbouring forest regions.

In Fig. 6.17 an attempt has been made to classify geographical regions with special reference to agricultural factors.

The coastal area (I) is subdivided into three main regions (Cape Coast-Winneba Plains, Accra Plains, Trans-Volta Plains). The narrow Nzima coast, with its immediate hinterland, has been included in the forest belt, although in part it resembles the coastal fringe in its economic structure. North of the thin coastal fringe, the thinly populated western forests and the main cocoa zone constitute the greatest expanse of the forest belt (II). The slightly drier, more densely populated southeastern area, transitional to the coastal savanna, is as important for the cultivation of field crops as is the Krobo region. The Mampong-Kwahu plateau, stretching north of the cuesta of the upper sandstone formation, is also more important for food crop cultivation than for cocoa-growing. Farther east, the northern hill country of the Akwapim-Volta is more important for cocoa-farming, whereas in the southern section rice cultivation predominates.

This classification of Ghana differs from others (e.g. Boateng, 1959; White, 1958) mainly because of the exclusion of the thinly-populated middle belt (III), which contrasts strongly with the more densely populated northwest and the Tamale-Yendi district. This middle belt comprises northwest Ashanti, the Afram Plains, Krachi, southern Dagomba and Gonja. Its southern border, along the dense forest belt, is quite clearly marked, but it is more difficult to draw a definite line in the north. The only suitable boundaries might be the southern edge of the slightly more densely populated Lawra-Wa district and the more intensively cultivated region along the Tamale-Yendi axis (III).

To sum up: Ghana may be considered to be an example of a political unit artificially created within rigid boundaries during the nineteenth-century colonial period, in which the natural geographical division into three regions—the coastal savannas, the forest belt and the northern savannas—determines its main agricultural geographical structure. This agricultural subdivision of Ghana is only gradually changing into a more strongly functionally determined subdivision under the influence of world-market orientated crop regions and urban concentrations.

The economy and habitats of primitive hunting and food-gathering groups

Sections of the Austral-Pacific culture zone have been chosen to illustrate the exploitation of natural regions by societies still in a primitive economic stage; early phases of development can still be traced in parts of this zone today, and they can provide us with information about man's development, although nowadays, ethnology no longer regards Australian aboriginal society as equivalent to the first stage of man's development.

Studies by prehistorians indicate that the Aborigines have been living in Australia for at least 10 000 to 15 000 years. At the time of the first extensive European colonisation in the nineteenth century, their numbers were still over 300 000, belonging to a few hundred different tribal groups. Population density varied between one inhabitant per 60–100 square kilometres in the 'dead heart' of the continent to one inhabitant per square kilometre on the coasts. These groups, who can be classified as belonging to the 'hunting and food-gathering' stage, were spread out over almost all the Australian continent; they lived in the most varied natural regions: mountain forests, savannas and deserts. In spite of this variety in their environments, the Aborigines did not develop a simple cultivation anywhere. They did not possess any domestic animals and their methods of preserving supplies were simple. Everywhere they lived from hunting, fishing and food-gathering, in the most varied combinations. Apart from barter they had hardly any economic relations which went further than self-sufficiency. Development progressed slightly faster on the coast than inland; besides a slow improvement of their tools and weapons, Austronesian influences can also be proved here. For example, the Macassar dugout canoe and the outrigger craft were adopted from Papua. But in the arid core region of Australia, the Aborigines had to cope with an environment comparable to that of the Bushmen of south-west Africa.

The habitats of the Australian Aborigines are classified according to the climatic and vegetation zones of the continent. There are three main habitats, which in some regions partly interlock:
1. the evergreen rainforests of the northeast; 2. deciduous woodlands and humid savannas, with their corresponding coastal and riverine forests; 3. the dry savannas and deserts of the interior.

Little is known about the original economy of the Australian aborigines of the northern rainforest region. It probably resembled that of the Negrito groups in Southeast Asia (the Philippines, Ceylon and Malaysia). The latter were mainly food-gatherers who lived on the fruits of the forest, with some additional hunting. They also created small forest fires during the dry season, but since the population density was so low the natural environment was hardly changed by this practice.

In the coastal region there was a sedentary population; there were also migratory groups who camped on the mangrove estuaries and lagoons.

The most important economic activities of these groups were fishing, carried out with the help of canoes, and catching turtles and crocodiles. Hunting on land was relegated to the background in these areas; only in some regions did the Aborigines hunt during the dry winter months, when they burnt off the dry grass. In spite of the importance of fishing, which was carried out by the men only, their food was mainly vegetable. The women of these coastal tribes gathered and used various tuberous plants, especially species of yams, which had to be prepared carefully. A large proportion of the edible plants could be gathered all the year round near the coast and near river banks and swamps. Often only parts of the tubers were cut off, so that the plants could sprout and the tubers grow again. They never progressed as far as simple horticulture, such as planting the tubers in suitable sites in a planned way—a step which seems so obvious to us. Religious beliefs are sometimes used as an explanation of this strange stagnation in economic development.

In the dry areas of the interior the Aborigines lived mainly from hunting. Apart from the 'channel country' in northwest Queensland the intermittent streams offered only poor prospects for fishing; but hunting with spears and boomerangs, and using fire and beaters, was incapable of feeding the tribes of the savannas and steppes. To a large extent the food supply was vegetable. Besides roots and fruits, grass and acacia seeds were collected, ground up by the women and made into simple flat cakes. In brief, it can be asserted that both among the coastal peoples and in the dry inland areas, vegetables constituted a large proportion (70 to 80 per cent) of the diet. In inland areas seeds (37 per cent), fruits (35 per cent) and tubers (11 per cent) were important components of this vegetable diet; on the coast, important constituents were fruits (45 per cent), root-tubers (27 per cent) and nuts and seeds (5 per cent each) (Megitt, 1963) However, this highly vegetable diet is not only the result of poor techniques among the inhabitants of Australia, but is also due to the scarcity of animals in the continent. Apart from the marsupials, which were few in number, there were no mammals in pre-European times which could be compared, for instance, with the large herds of game in Africa or the herds of bison in North America. The higher density of the Australian Aborigines on the coast is partly explained by their more nourishing vegetable diet, which was less limited by dry spells, and the better protein supply from fishing.

In contrast to these Australian game hunters, in New Guinea one can observe a distinctive horticulture of yams and batatas (sweet potatoes) in clearings made by burning, together with the planned exploitation of trees such as the sago palm. At least two cultural stages can be proved with certainty here. The Papuans constitute the older stage, which can perhaps

[1] This highly vegetable diet also seems to be typical of other hunting peoples who go in for food-gathering. Marshall (1960) estimates it at 80 per cent for the Kung Bushmen in South Africa.

be connected with the Aborigines of Australia, even if their cultural development took a different course; the Papuans remained at a low cultural level up to the time of the second large wave of migration. Presumably they were still pure hunters and food gatherers, or only knew the most primitive forms of cultivation. New Guinea owes its further cultural development to the large Austronesian migratory movements which swept over what are today Indonesia, the Philippines and wide areas of Oceania in several waves during the first and second millennia BC. The best known are the Austronesian groups who reached Tonga and Samoa in the first centuries before Christ; there they developed the Polynesian culture and discovered the Pacific archipelagoes in the course of daring voyages in their fast sailing craft. Other Austronesians, who also came from the west, settled on the coast of New Guinea. As representatives of a dynamic culture they did not stop there but began to penetrate into the interior of the island, mainly along the big rivers. The stimulus of these Austronesian 'bearers of civilisation' left behind a distinct legacy. Not only did it trigger off a further development of their total culture, with new political forms and religious ideas, it also contributed to the birth of a New Guinea art whose creations are among the most beautiful examples of primitive art in existence. Since the Austronesians brought with them the technique of utilising the wild sago palm tree, swamps so far uninhabited could now be settled. The introduction and cultivation of yam tubers led to a higher standard of tillage, and in intensifying the growing of batatas the population of the central plateau was increased to an extent which is amazing in view of New Guinea conditions (Haberland, 1964).

Similar types of culture can be observed in the areas of 'retreat' or isolation in the rainforests and wet forests of Amazonia, Southeast Asia and tropical Africa. They are, however, characterised by a marked fragmentation into small ethnic and linguistic tribal groups. Often the principal axes of development were the rivers, along which a certain amount of contact existed through barter and trade. Fishing was more important than hunting. Groups like the Semang and Senoi, living on the Malay peninsula, had already progressed to the stage of harvesting tubers and roots with simple digging-sticks. They also gathered various fruits and the seeds of the Perah tree. In order to avoid the seeds being blown too far away they topped the trees and cleared the nearer surroundings of undergrowth; these are measures which can be regarded as a first stage in arboriculture and weed-killing. In Black Africa, as well as the pygmies, the game-hunting tribes of Bushmen in southwest Africa and the Kalahari are also classified as belonging to a lowly cultural stage; they formerly inhabited a very wide area.

In contrast to the pygmies of the Congo rainforest or the corresponding groups in Amazonia, the old Australian hunting, fishing and food-gathering peoples reshaped the arid regions of their environment considerably by using fire for hunting and for fighting off snakes. Areas of many square kilometres can be devoured by flames within a few hours in the dry season, and may be damaged permanently by wind erosion or rain later on. The large number of fire-resistant trees is a pointer in this direction.

It is very difficult to reconstruct the areal extent of former phases of agricultural development, as extensive parts of northern Australia have been transformed by European colonists who practised a pastoral economy and who set up large cattle and sheep farms. In this connection, the contrast between the very dynamic subtropical agricultural development of Australia and the 'empty north' is noteworthy. Outside the sugar plantations in eastern Queensland there is very little intensive agricultural cropping in the Australian tropics. (One exception is the Ord river basin, see Fautz, 1970.) The isolated station that, in the rainy season, can only be reached by plane and where long droughts as well as floods may lead to serious losses of cattle, is still typical of this region.

The transformation of tropical environments by agriculture

Any attempt can now be made to provide an overall view of the differing extent to which agriculture has contributed to the remarkable changes of tropical environments.

Regions can be graded into different degrees of transformation, ranging from the hardly altered virgin forest to the completely transformed cultural landscape, according to population structure and distribution, economic stage and type of land use. In the more densely populated forest and savanna areas the original natural conditions have often been considerably transformed. In most of the humid forests and rainforests it is frequently difficult to state whether they represent primary forms or not. Primary or secondary vegetations can only be distinguished with the help of plant sociology methods of investigation.

The wholesale destruction of tropical forests in Africa, Asia and Latin America has reached the proportions of a major ecological crisis. While for a long time it was difficult to get a clear picture of the extent of these losses, in recent years botanists, silviculturists and geographers have obtained more reliable background material concerning this threat, which obviously has serious consequences not only for the soils, the water balance and the climate of the tropical zone, but also for more global changes of the environment.

It appears that the greatest losses have occurred in West Africa and Southeast Asia. The area of the African rainforest which was estimated to be about 218 million hectares has shrunk to about 60 per cent of its original size. It is estimated that between 1930 and 1970 alone, about 25 to 30 per cent of the African rainforest was destroyed. In the coastal and densely populated areas, the proportion of forest that has been transformed into secondary bush is even higher. In Sierra Leone, for instance, the forest area which previously covered nearly two-thirds of the country has dropped to about 5 per cent. Also in Liberia, the area under rainforest was reduced within a few decades from 9·3 million ha to 3·6 million ha; for several years over 20 000 ha were cleared for the growing of hill rice in Liberia. In Southwest Ghana, of the original area of 8·2 million ha, about 5·8 million ha were changed into farmland and bush fallow. The annual loss is here estimated at 52 000 ha. In spite of government action and the establishment of forest reserves, the destruction of non-reserve forest proceeds rapidly. Hesmer (1966) believes that by 1980, only 18 per cent of the original forest in Ghana will be left. By the end of the century, almost all rainforest in the West African forest belt between Sierra Leone and Ghana will be changed into secondary formations, bush fallow and cultivated land. Even in the large forest areas of the Congo Basin, the extent of forest destruction is serious, and losses are estimated to be around 50 per cent.

In south and southeast Asia, the annual losses through forest clearing amount to over 15 million ha (Indonesia, 5·26 million, India 4·8 million, and Burma 1·15 million, and the Philippines 0·73 million). Only some 4·5 million ha have been reafforested. In addition, the use of herbicides in the Vietnam war has negatively affected over 2 million ha of rainforest. The annual forest losses in Latin America were estimated by the FAO to be about 10 million ha.

The main causes for this large-scale disruption of tropical vegetation lie in the practice of slash-and-burn cultivation by tropical peasant groups. The acquiring of larger plantations and the commercial exploitation of timber resources has a lesser importance. The ecological consequences of these destructive processes are manifold: soil erosion, a marked change of the vegetation formations, changes of sedimentation, runoff and other hydrological facts seriously influence the whole tropical environment.

Numerous studies have investigated the questions of climatic fluctuations and especially the progressive desiccation of some tropical regions. Historical sources, or accounts written by early explorers, provide valuable clues about changes in the environment; such changes can have conflicting causes, and lively scientific discussions have been engendered in this way. Since the periods for which exact climatological evidence exists are relatively short, many existing studies or theories about long-term changes in climate, soils and vegetation have to be treated with caution. Even the frequently quoted progressive desiccation of tropical Africa (Stebbing, 1935; Jaeger, 1943b; Kollmannsperger, 1965), and the often postulated southward expansion of the Sahara are not accepted by all authors. Although important long-term changes have undoubtedly taken place— there is much evidence for this in rock drawings and traces of old settlements—the causes have not yet been examined sufficiently. Apart from worldwide changes since the ice ages, or since the period of greater rainfall during the Pleistocene—changes which are probably due to exogenic (actually solar) influences—great importance has to be attributed to

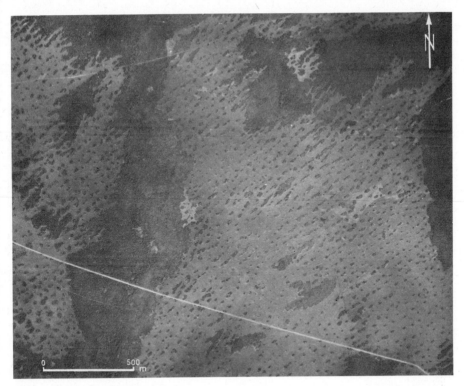

PLATE 15. Degradation of forest vegetation (Winneba Plains, Ghana). Thicket growth
is here transformed by fire into open savanna country. The influence of the
predominant southwesterly winds is clearly recognisable. Air photo 14.3.1954.

anthropogenic effects, and it is these that will be discussed here.

In Africa the groundwater level has become lower over extensive areas
(Harroy, 1949). Large tracts of rainforest have been degraded and transformed into wet savannas; wet savannas have turned into dry savannas;
and dry and thornbush savannas have become semideserts. Over the centuries, lack of knowledge about natural interrelationships, for example
between climate, soils and vegetation, have created manmade deserts with
their manifold symptoms of devastation, such as incrustation, soil exhaustion and leaching. These manmade deserts were extended by overgrazing
and the random clearance or burning-off of the vegetation. Destructive
exploitation of the land in the tropics goes a long way back. The great
population increases of the nineteenth and twentieth centuries and intensified economic production for home and world markets during the colonial
period, have accelerated these processes.

Under the influence of such long-term anthropogenic factors as shifting
cultivation and grazing, a slow regression of forest boundaries can be
observed in the transitional or contact zones between those forms of vegetation in which ecological instability is typical (Morgan, Moss, 1965). The

degradation of tropical vegetation has been examined thoroughly in Madagascar, for instance, where a large part of the forest was destroyed or degraded to *savoka* secondary types (Humbert, 1927; Lavauden, 1931). On this island, which is free of tsetse fly and which is also known as *Isle de Boeuf*, the breeding of Zebu cattle played a considerable role in this process during the last few hundred years. In other tropical areas this kind of development was accelerated by intensive shifting cultivation. The secondary grasslands in such countries (e.g. Cogon or Alang-Alang grasses, *Imperata cylindrica*, in Indonesia, Malaysia and the Philippines) are similar in both their physiognomy and their lack of nourishment. Another widespread type of secondary vegetation is found in the various kinds of thicket which grow in many parts of the tropics. The degree to which human influences made themselves felt in each individual case is still disputed.

Right up to the present, a great deal of attention has been paid to the 'savanna problem'. It is generally accepted that the existence of savannas in the humid tropics is hardly ever caused by climate alone. It can be regarded as certain that many tropical savannas are of manmade origin and were formerly forests. There are types of savannas which may be derived from edaphic factors, although it is still disputed how they were formed and to what extent man influenced their formation. Anthropogenic factors must therefore be held mainly responsible for many savanna ecosystems ('induced' or 'derived' savannas). However, besides the very obvious dependence of vegetation formations on climate, a more intensive study should also be made of geomorphological and edaphic factors such as relief, soils and erosion.

It is perhaps a matter for surprise that in many parts of the globe, and especially in the New World, savannas were used more for animal husbandry and grazing than for cultivation. The main reason for this is the fact that the loose soil of the forest, cleared by fire, can usually be worked more easily and is more fertile (thanks to the large amount of ash it contains) than the soil of the savanna, on which the grass soon springs up green again after surface burning. Only with the help of modern methods, partly introduced by European colonial governments, is it possible to open up grasslands to intensive agriculture. On the other hand, typical grassland associations soon begin to establish themselves in the cleared 'islands' in secondary forests.

Ecological consequences of traditional farming systems, such as shifting cultivation or slash and burn techniques have been considerable, although in previous centuries man has developed a kind of common sense towards the exploitation of natural resources. The long-range environmental impact (e.g. in the transformation of forest into secondary bush) has been much greater than was formerly thought. These changes have become increasingly evident with growing population densities.

There exists a marked imbalance between tropical peoples and their life support systems which can partly be explained by population densities.

But this imbalance can also be caused by agricultural systems imported from outside which are 'ecologically inappropriate and socio-economically prejudicial' (Dickinson, 1972, p. 713). Often high quality land is used for export crops and cattle-rearing. This may lead in turn to the displacement of traditional agriculture which can trigger off increased migration into towns or the further destruction of marginal lands.

Perhaps more efficient agricultural management can be achieved by simulating structure and diversity of tropical ecosystems as alternatives to the monocultural systems that have been introduced in the tropics. Some of these new systems reach their highest productivity levels only at considerable cost to the natural environment.

Outlook

Some aspects of the 'Green Revolution'

An interesting feature of recent agricultural development in tropical and subtropical countries is the much publicised 'Green Revolution'. This was connected with the breeding of new high yielding varieties of wheat (in Mexico) and rice (in the Philippines). Within the last decades, these new strains have been successfully introduced into many other tropical environments—though in tropical Africa the revolution has so far had no important results. This new trend led to some hopeful technological innovations in farming communities that could considerably increase their cash income. Growing prosperity in some countries has led to a declining dependence on food imports, and also to a stimulation of the non-farming sector of the economy.

Thus, the Green Revolution must primarily be considered as an agricultural technology using high-quality seeds, chemical fertilisers, pest control, irrigation and other modern land use methods which lead to an increase of crops per area sown. The new varieties require costly irrigation and carefully controlled lands. Another question is how far climatic and weather changes partly account for the success of the 'miracle seeds'. In India, for instance, the great improvement in the self-sufficiency of food crops was supported between 1966 and 1971 by good monsoonal seasons. A serious setback was experienced after the drought in 1972, however, which underlined the fragility of the achievements by the Green Revolution (Curry-Lindahl, 1972). Further, because of its enormous demands on chemical fertilizers, there is no doubt that ecologically the Green Revolution endangers the biosphere.

Much of the most suitable land has now been cultivated with these high-yielding strains. The 'cheap' phase of the development seems to be over. For further expansion great investments in irrigation and land reclamation will be necessary.

In connection with some over-optimistic appraisals other problems connected with cereal self-sufficiency must be borne in mind. To a starving man cereals are basic. But as the man's income rises above starvation level, the role of cereals begin to shrink, the diet changes, and in many tropical countries diets are changing at a considerable pace. The growing demand for protein is now becoming a world problem. But the dangerous protein gap cannot easily be closed with these methods. Nutritional problems have, in the first place, to be closely linked to population control and economic organisation, because feeding mankind is not only a problem of production, but more so of marketing and distribution. It is obvious that food production cannot be increased indefinitely. The problem is less that of a food shortage alone than of too many human beings.

For the man who has the money new agricultural techniques are interesting. He can buy fertilisers, pesticides and modern equipment. He can hire extra labour. For all this, the richer farmer is better placed. Therefore these techniques are tending to create new social groupings with new income gaps in rural areas.

Irrigation schemes that stop flooding and alleviate drought are important, but most of them do not provide for better water management. For higher yields and double-cropping the proper level of water at the right time is required. If too much water goes into the field, all the expensive fertilisers may be washed away.

The unexpected increases of agricultural production have led in many regions to bottlenecks in transport, storage and marketing facilities. It has also created problems with such productive factors as water, fertilisers or credits. If the new potential of the Green Revolution is to be used fully, drastic reforms in land tenure, demographic control as well as new forms of social and economic behaviour are required.

The breakthrough in cereal production has brought about new conditions in the fight against hunger. Previously, it was considered to be very unlikely that the additional food needed in poor countries could be produced by themselves. In spite of largely illiterate and geographically widely dispersed farming communities, this progress was achieved and in some cases even a surplus could be exported. 'The problem now is to keep agriculture moving. Otherwise, the farm breakthrough will aggravate the job shortage and accelerate the exodus from the countryside to the already overcrowded cities' (Brown, 1970, p. 195).

In briefly summing up the main problems of the Green Revolution the following points may be made:

SUCCESSES: The Green Revolution has in some countries led to remarkable increases in cereal production, especially of wheat (cf. Table opposite).
PROBLEMS: The chances of a further considerable extension, with growth rates equal to those of the early years, are not too good. Without birth control the Green Revolution cannot bring about a long term improvement

		Production of wheat, rice and maize (in million tonnes) :			
COUNTRY	CEREAL	1966/67	1968/69	1970/71	1971/72
India	Wheat	11·39	18·65	23·25	25·50
	Rice	30·44	39·76	42·45	38·00
	Maize	4·89	5·70	7·41	not available
Pakistan	Wheat	3·85	6·51	6·37	6·83
	Rice	1·30	2·00	2·16	2·40
	Maize	0·53	0·62	7·06	6·94
	Consumption of fertilisers				
India	Nitrogen	740	1 200	1 479	1 810
	Phosphate	200	410	540	590
	Potash	116	170	236	350
Pakistan	Nitrogen	112	205	251	343
	Phosphate	4	40	30	38
	Potash	—	2	1	1

Source. Mullick, 1973.

of the economic situation. It cannot bridge the protein gap. Nutrition and labour problems are closely interconnected. Food supply is not only a matter of production but also of distribution and marketing.

DANGERS: The Green Revolution, through its excessive use of chemical fertilisers and pesticides, may endanger the environment of tropical countries. Income gaps tend to be accentuated. Political and trade conflicts on the world market as well as on the local markets may result.

Planning the future

An important problem for the future of agricultural planning in the tropics is the maintenance and regeneration of tree stocks. Besides the traditional forest reserves of the native population—which were limited in area and whose exploitation was forbidden on religious grounds ('holy groves') the colonial powers introduced protective measures for natural vegetation. In doing so, they often met with fierce resistance from the farmers. In recent years, however, the governments of developing countries in the tropics have realised the value and importance of forest reserves in terms of climate, groundwater, soils and the regulation and restoration of the balance of nature.

Shifting agriculture and land rotation played an important part in the agricultural development of the tropics. However, they can and must be superseded one day, as new techniques and financial resources become available. Under previous, and sometimes even present, conditions of relatively feeble population pressure, large reserves of land and traditional forms of social organisation, these land use systems were (and still are) well suited to the simple needs of the community and the ecological equilibrium. It is only with an increase in population, due to improved conditions of hygiene and dwindling land reserves, that these practices become a difficult and still largely unsolved problem, calling for an improvement of and change in agricultural methods. These will necessitate greater attention being paid, in future, to regional variations in climatic conditions. In wet forest climates, for instance, vegetation renewal takes place much faster after cultivation than in savanna or arid regions. The extent of soil – erosion is also closely connected with vegetation cover; and, finally, the degree to which leaching of nutritive minerals takes place differs between forests and savannas. It has really become a matter of urgency to replace these farming systems, which are so solidly integrated into their geographical surroundings, if economic and human resources are to be freed for the industrialisation of the tropics—a process which is slowly beginning.

There are many different starting points for solving these problems. In India, for example, great interest is being shown at present in large-scale water conservation and irrigation projects (e.g. the Ganges Water Grid); these projects also often supply power for industry. Such undertakings reduce the present dependence of crop yields on climate; the periodic famines which occur in densely populated areas when the very variable rains fail to come on time or in sufficient quantity are evidence of the closeness of this dependence. It is not enough, however, to reduce natural hazards; at the same time, agrarian reforms and new agricultural techniques must be introduced on a large scale, to relieve existing difficulties.

In the many parts of tropical Africa (Ghana and Uganda, among others) where so far cash crops play a greater part, the supplying of local markets should be used as a pivot for agricultural development. Such a policy would take into account both the increasing urbanisation of these areas and the question of a better food supply for the urban population. As the symbiosis between town and country becomes ever closer, a better use of urban waste products would become important (e.g. for market gardening).

Such innovations originate in certain core areas with modern economies, and spread into the surrounding regions. The need arises to change traditional social customs and ways of life, a need which in many tropical countries entails so many difficulties that it may almost appear to be a labour of Sisyphus. New methods of communication have to be found which will avoid too strong a clash between progressive and traditional ways of life and will achieve permanent success. The difficulty rests in the fact that social customs, deeply rooted in tradition, obstruct peoples' readiness to accept innovations. Ways have to be found of exerting a positive influence, through training and education, and rendering possible more rational ways of life. One of the key issues in these countries is the solution of land tenure problems. One question in particular presents itself more and more acutely: whether, in order to build up a viable national economy, to allow the growth of large enterprises or whether to share out the land among poor tenant farmers and labourers, for social reasons. The governments of developing countries are frequently accused of being too socially minded instead of thinking in economic terms. There are, in fact, many areas which are no longer worth cultivating after a land reform, for lack of suitable tools and manuring; roads, paths and bridges which were formerly maintained by landowners and large farmers fall into decay.

It seems questionable whether economic progress can be promoted by comprehensive development programmes initiated by governments alone. In the final analysis, economic efficiency is always determined by the qualities, attitudes and drive of certain human groups who are the pillars of their societies. Differences in economic strength nearly always reflect different social and institutional backgrounds, and if the prerequisites mentioned above are missing, the tropical countries concerned can hardly hope for a rapid leap forward in development. It is often more important, therefore, to stimulate intellectual resources through education, than to invest large amounts of development aid in the hope of rapid industrialisation.

It is difficult to delineate an exact picture of the agricultural geography of the tropics. Some of the examples and conditions which have been described will certainly not be relevant any longer, in a few years' time, because it is precisely in the tropics that changes are taking place extremely quickly. Many progressive farmers are now daring to try new methods. Religious taboos are increasingly cast aside and overcome. With the help of foreign aid, governments develop model farms or set up cooperatives to serve as examples for the general development of the community.

The small successes achieved through the improvement of agriculture and the modernisation of the economy have not kept pace with the rapidly rising population graph (India's rate of increase is over 10 million a year) so that the prospects of maintaining the population worsen or at best stagnate. Many relief measures are piecemeal, and political barriers often prevent positive developments.

Since geography is the only science which is directly concerned with spatial relationships, and which takes such great interest in the study of economic, social and cultural interactions, it lends itself naturally to the coordination of aid programmes and development projects. Sandner (1966, pp. 74–6) formulated some of the tasks awaiting applied geography in the field of regional planning, and grouped them into a series of five work stages:

1. The prerequisite for all practical and planning activity is an exact

knowledge of what has existed hitherto; generally speaking, this implies not only geographical situation or structure, but equally the style of development within a country, its dynamics, and an understanding of the principles governing the types of spatial patterns to be found there. It is not so much a matter of making a summary of all our knowledge of a country, or an inventory of all the facts which might be useful to the geographer, but rather coming to understand its principal distinguishing characteristics. Careful regional planning, for instance, demands a very detailed knowledge of geographically significant behaviour patterns within the country's population. Embraced within this field of knowledge is not only the comprehension of economic attitudes, living styles and social behaviour, but also the kind of spatial patterns which are favoured, and which become apparent, for instance, in new settlement zones, if development is allowed to take place freely and unchecked. This first stage of work comprises two processes: on the one hand a geographical stocktaking of the planning area in question, with the help of air photo interpretation, mapping of the countryside, and questionnaires; on the other hand, the geographical characteristics and qualities of development have to be worked out over a large area—an equally important and far more difficult kind of study.

2. In the second stage of work the development possibilities of the planning area are studied. First of all an individual analysis is made of soil qualities and their suitability for different crops, their cultivation capacity, the suitability of the climate for different ethnic groups, the possibilities of creating integrated markets, and other phenomena. In this way one starts off with a very wide range of development possibilities and one has to resort to seeking the help of special study groups.

3. The planning proposal is studied by means of discussions between the different contributors and departments; in the course of these discussions the plan developed by the geographer is confronted by ideas which have a stronger political or technical orientation—views which contrast with those of the geographer, which are more related to the spatial dimension and which try to integrate phenomena.

4. The next step—that of finding a way to carry out the plan and bring it down to the level of concrete realisations—no longer really belongs to the geographer's sphere of activity, although his collaboration is a prerequisite for it.

5. This section of activity, which is very fruitful from the scientific point of view, has often been neglected in the planning process. Careful observation of an area which is experiencing development planning serves the purpose, not so much of rectifying one's own ideas about how to prepare such projects, as of contributing to scientific knowledge. To this end, one needs to follow up the projects as they proceed, and to study such features as labour demands, economic yields and social attitudes.

The way in which it is defined above shows that 'applied geography' is regarded neither as an independent branch of the subject nor as the anti-thesis of scientific geography; it is shown to be a geographical activity which is defined only by its individual form, which aims at attaining practical goals and efficiency; this kind of activity reflects the great variety of starting points, views and ways of asking questions which is typical of our subject.

Another good example of the geographer's contribution to a social survey in a tropical country is Fortes, Steel and Ady *Ashanti Survey 1945–46*.

In the new political phase of development, wherein the former trans-regional and main colonial spatial patterns have given way to an introverted nationalism, in most cases there are good grounds for putting comparative geographical ways of thinking into the forefront. In the African case, some of the prophecies made by Dumont in his controversial book written in 1962 *L'Afrique noire est mal partie* (English translation *False start in Africa*, 1966) have not only come true, but the economic situation in many respects is worse than in colonial times. In Black Africa annual food production has increased by only 1·7 per cent since 1959, while the population has grown by 2·5 per cent per annum. In demographic studies this phenomenon is often referred to as the 'low level equilibrium population trap' (cf. also Hodder 1968).

It is obvious from reports of the Economic Commission for Africa (E.C.A.) that agriculture has faced great difficulties in feeding the rapidly-increasing population and that it has been very hard to provide newly-created industries with raw materials and sufficient foreign exchange.

About one-fifth of the population of development countries are hungry and under-fed, two-thirds suffer from various degrees of malnutrition. Protein deficiency for young children is a grave problem with severe consequences for their mental development. Out of the total number of the

Overall index numbers of needs in total food and animal food supplies in A.D. *2000 under long-term target, by regions (1957–59 = 100)*

REGION	TOTAL FOODS	ANIMAL FOODS
Far East	417	654
Near East	334	463
Africa	337	533
Latin America (excl. River Plate countries)	359	403
Low-calorie countries	393	583
World	274	308

Source. Third World Food Survey, FAO, Rome, 1963, cited by Grigg (1970) and Wright (1972); cf. also Boesch (1972).

Total world population, 1965, and projections to 1980

REGION	1965	1980	PROJECTED ANNUAL RATE OF GROWTH
	Thousands		*Per cent*
Developed countries			
United States	194 572	235 200	1·4
Canada	19 604	26 024	1·9
EEC	181 594	198 385	0·6
United Kingdom	54 595	60 690	0·7
Other Western Europe	87 684	97 489	0·7
Japan	97 960	111 563	0·8
Australia and New Zealand	14 000	18 216	1·8
South Africa, Republic of	17 867	26 000	2·6
Total developed	667 876	779 446	1·0
Central plan			
Eastern Europe	121 430	138 763	0·9
USSR	230 600	277 325	1·3
Communist Asia	795 604	1 077 064	2·0
Total central plan	1 147 634	1 493 152	1·8
Less developed			
Mexico, Central America, and Caribbean	80 078	128 508	3·2
South America	166 046	247 185	2·7
East and West Africa	217 454	315 620	2·5
North Africa and west Asia	162 483	254 032	3·0
South Asia	638 064	913 655	2·4
Southeast Asia	81 057	117 969	2·5
East Asia and Pacific Islands	198 597	298 920	2·8
Total less developed	1 543 779	2 275 889	2·6
WORLD TOTAL	3 359 289	4 548 487	2·0

Source. US Department of Agriculture, 1970.

world's 950 million children, about 600 million are not fed properly. In order to change these conditions, the total food production in some tropical countries until the year 2000 has to be quadrupled, the production of animal food products would have to be increased even more.

The results of the First Development Decade, started by the United Nations in 1961, have been disappointing. In fact, in many cases development has slowed down in comparison with the 1950s. The contrast between the poor and the rich countries, that is, between the predominantly tropical countries and the industrial countries, has steadily increased. The approximately twenty industrial countries of Europe, North America and Japan which contain fewer than one-fifth of the world's population, produce and enjoy more than half of the earth's riches. In contrast, those developing countries affiliated to the World Bank contain half of the world's population but have to be satisfied with one-sixth of the world's production.

There is no doubt that unstable political conditions are partly responsible for these circumstances in many developing countries. It is difficult to maintain per capita income when the mean annual population increase is 2·5 per cent or more. Economic progress is constantly swallowed up by the increase in population. Typically, it is agricultural production in particular which is showing the slowest increase. The better development performance of industry and mining is partly due to their receipt of development aid. Such aid, however, is far more difficult to apply efficiently to agriculture. The agricultural sciences and geography can at least offer some help in this vast field.

Since the underfed or badly nourished two-thirds of mankind live in the tropics or subtropics, and since the frequently cited population explosion is taking place precisely in these regions, there is an urgent need to carry out analyses of the structures of the developing countries; this task cannot be solved from the points of view of economic and social scientists alone. In future, regional reports, ranging from studies of basic ecological or social regions to monographs on entire countries, should make more reference to geographical parallels or analogues. Only in this way can knowledge gained about one area be extended with some degree of efficiency to another.[1]

Summary of conclusions

In comparison with the industrial countries agricultural productivity in many tropical countries is very low. There is a lack of technical and scientific know-how, of equipment, of fertilisers and pesticides. The infrastructure for communication, for marketing and distribution is inadequate. Although a more rapid industrialisation is important to trigger the economic take-off, it seems that a main impulse must come from agricultural development.

All modernisation of tropical agriculture is closely tied up with the balanced development of industry and commerce. An important prerequisite for modernisation is the establishment of better systems of land tenure. Without radical land reforms, all attempts for a better feeding of

[1] Practical development aid has made a successful start in this direction, e.g. by sending Chinese agricultural experts to tropical countries.

the world's population are illusory. Much of the present instability stems from this unequal distribution of land. In many ways the simple techniques of traditional tropical agriculture reflect the social, political and religious situation, and a state of mind of populations who, in a few generations, aim at reaching a level of development that has taken the western world several centuries.

Bibliography

ABEYWICKRAMA, B. A. (1963) *Pre-Industrial Man in the Tropical Environment*, International Union for the Conservation of Nature and Natural Resources, Ninth Technical Meeting, Nairobi.

ADEGBOYE, R. O. (1971) 'Customary land tenure systems in Uganda', *Land Reform*, FAO, Rome, **1**, 68–73.

ADICEAM, E. (1966) *La Géographie de l'irrigation dans le Tamilnad*, Paris.

AEREBOE, F. (1920) *Allgemeine landwirtschaftliche Betriebslehre*, Berlin.

AGBOOLA, S. A. (1972) 'Agricultural typology in Nigeria—problems and prospects', *International Geography* (papers submitted to Int. Geog. Congress, Montreal, Toronto, 1087–8.

AHMAD, NAFIS (1961) 'Soil salinity in West Pakistan and means to deal with it', in *Salinity Problems in the Arid Zones: Proc. Teheran Symposium*, Unesco, 117–25.

ALBRECHT, H. (1972) 'The concept of subsistence', in *Zeitschrift für ausländische Landwirtschaft*, Frankfurt, 274–88.

ALEXANDER, J. W. (1963) *Economic Geography*, Englewood Cliffs, Prentice-Hall.

ALLAN, W. (1948) *Land Holdings and Land Usage among the Plateau Tonga of Mazabuke District*, Kapstadt.

ALLAN, W. (1960) 'Changing patterns of African land-use', *Journal of the Royal Society of Arts*, London, **108**, 612–29.

ALLAN, W. (1965) *The African Husbandman*, London, Oliver & Boyd.

ALLISOW, B. P. (1954) *Die Klimate der Erde*, Berlin.

ANDREAE, B. (1964) *Betriebsformen in der Landwirtschaft*, Stuttgart.

ANDREAE, B. (1966) *Weidewirtschaft im südlichen Afrika. Standorts- und evolutionstheoretische Studien zur Agrargeographie und Agrarökonomie der Tropen und Subtropen*, Wiesbaden.

AUGÉ, M. (1969) 'Le rivage Alladian—organisation et évolution des villages Alladian', in *Mem. ORSTOM*, Paris, no. 34.

BAADE, F. (1964) 'Entwicklungsstrategie für die Weltwirtschaft von morgen', in *Der Volkswirt*, 58–63.

BAKER, O. E. (1926) 'Agricultural regions of North America', *Econ. Geogr.*, **2**, 459–93.

BAKER, S. (1966) 'The utility of tropical regional studies', *The Professional Geographer*, **18**, no. 1, 20–2.

BALOGH, T. (1966) *The Economics of Poverty*, London, Weidenfeld & Nicolson.

➤ BARTLETT, H. H. (1956) 'Fire, primitive agriculture, and grazing in the tropics', Thomas (1956), 692–720.

BARTZ, F. (1957) 'Die Insel Ceylon: Gesellschaft, Wirtschaft und Kulturlandschaft', *Erdkunde*, **11**, 249–66.

BATES, M. (1952) *Where Winter Never Comes*, New York, London, Gollancz.

BAUER, P. T. (1963) *The Study of Underdeveloped Economies*, London, London School of Economics.

BAUER, P. T. and YAMEY, B. S. (1957) *The Economics of Underdeveloped Countries*, Cambridge, Nisbet.

BEAUJEU-GARNIER (1972) *Images economiques du monde*, Paris (annual).

BEGUIN, H. (1960) *La Mise en valeur agricole de sud-est du Kasai*, Inst. Nat. pour l'étude agronomique du Congo, Sér. scientifique no. 88, Brussels.

BEHRENDT, R. F. (1960) 'Stichwort "Entwicklungsländer"', in *Handwörterbuch der Sozialwissenschaften*, 232–41.

BEHRENDT, R. F. (1965) *Soziale Strategie für Entwicklungsländer*, Frankfurt.

BELSHAW, D. G. R. (1963) 'An outline of resettlement policy', in *Uganda 1945–1963*, East African Inst. Soc. Res. Conf., June.

BENNEH, G. (1972) 'The response of farmers in northern Ghana to the introduction of mixed farming: a case study', *Géogr. Annaler*, 95–103.

BENNETT, M. K. (1954) *The World's Food*, New York, Harper.

BERGMANN, TH. (1966) 'Stand und Formen der Mechanisierung der Landwirtschaft in den asiatischen Ländern', Stuttgart.

BERRY, B. J. L. (1960) 'An inductive approach to the regionalization of economic development', in Ginsberg (1960), 78–107.

BERRY, B. J. L. and RAO, V. L. S. P. (1968) 'Urban-rural duality in the regional structure of Andrah Pradesh: a challenge to regional planning and development', *Geogr. Zeitschr.*

BEUKERING, J. A. VAN (1947) 'Het ladang-vraagstuk, een bedrijfsen sociaal economisch probleem', *Mededeelingen van het Department van Economische Zaken, Batavia*, no. 9.

BIEBUYCK, D. (1963) *African Agrarian Systems*, London, Oxford University Press.

BIEHL, M. (1966) *Die Landwirtschaft in China und Indien*, Frankfurt/M.

BLANCKENBURG, P. VON (1965) 'Afrikanische Bauernwirtschaften auf dem Weg in eine moderne Landwirtschaft', *Zeitschrift für Ausländische Landwirtschaft*, Frankfurt, **3**, 111.

BLANCKENBURG, P. VON and CREMER, H. D. (1967) *Handbuch der Landwirtschaft und Ernährung in den Entwicklungsländern*, Stuttgart, vol. 1 and 2.

BLAUT, J. M. (1953) 'The economic geography of a one-acre farm in Singapore', *Malayan Journ. Trop. Geogr.*, **1**, 37–48.

BLAUT, J. M. (1959) 'Microgeographic Sampling: a quantitative approach to regional agricultural geography', *Econ. Geogr.* **35**, 79–88.

BLAUT, J. M. (1961) 'The ecology of tropical farming systems', *Revista Geografica*, **28**, 47–67.

BLOHM, G. (1948; 2nd edn., 1959) *Angewandte landwirtschaftliche Betriebslehre*, Stuttgart.

BLOHM, W. (1931) *Die Nyamwezi—Land und Wirtschaft*, Hamburg.

BLUME, H. (1961) 'Die britischen Inseln über dem Winde (Kleine Antillen)', *Erdkunde* **15**, 265–87.

BLUME, H. (1961) 'Die gegenwärtige Wandlung in der Verbreitung von Groß-

und Kleinbetrieben auf den Großen Antillen', in *Schriften Geogr. Inst. Univ. Kiel* **20**, 75–123.

BLUME, H. (1964) 'Die Versalzung und Versumpfung der pakistanischen Indusebene', *Schriften des Geogr. Inst. der Univ. Kiel*, **23**, 227–45.

BLUME, H. (1968) *Die Westindischen Inseln*, Braunschweig.

BLYN, G. (1961) 'Controversial views on the geography of nutrition', *Econ. Geogr.*, **37**, 72–4.

BOATENG, E. A. (1959; 2nd edn. 1966) *A Geography of Ghana*, Cambridge University Press.

BOATENG, E. A. (1962) 'Land-use and population in the forest zone of Ghana', *Bull. Ghana Geogr. Ass.*, Accra, **7**, no. 1–2, 14–20.

BOBEK, H. (1948) 'Stellung und Bedeutung der Agrargeographie', *Erdkunde*, **2**, 118–25.

BOBEK, H. (1959) 'Die Hauptstufen der Gesellschafts- und Wirtschaftsentfaltung in geographischer Sicht', *Die Erde*, **90**, 259–98.

BOBEK, H. (1961) 'Über den Einbau der sozialgeographischen Betrachtungsweise in die Kulturgeographie', in *Dt. Geographentag*, Köln 1961, 148–65.

BOBEK, H. (1962a) *Iran: Probleme eines unterentwickelten Landes alter Kultur*, Frankfurt-Berlin-Braunschweig.

BOBEK, H. (1962b) 'Zur Problematik der unterentwickelten Länder', *Mitt. Österreich. Geogr. Ges.*, **104**, 1–24.

BOESCH, H. (1947) *Die Wirtschaftslandschaften der Erde*, Zürich.

BOESCH, H. (1962) 'Bewässerungsprobleme in West-Pakistan', *Geogr. Helvetica*, **17**, 222–9.

BOESCH, H. (1964) *A Geography of World Economy*, New York, Van Nostrand Reinhold.

BOESCH, H. (1965, 1966) 'Vier Karten zum Problem der globalen Produktion', *Geogr. Rundschau*, **17**, 303–16; **18**, 81–5.

BOESCH, H. and STÄDELI, A. (1968) 'Karten zum Problem der globalen Produktion', *Geogr. Rundschau*, **20**, 10–11.

BOESCH, H. and BÜHLER, J. (1972) 'Eine Karte der Welternährung', *Geogr. Rundschau*, **24**, 81–2.

BOHANNAN, P. (1954) *Tiv Farm and Settlement*, London.

BOHANNAN, P. (1963) ' "Land", "tenure" and land-tenure', in Biebuyck (1963).

BOHANNAN, L. and BOHANNAN, P. (1953) *The Tiv of Central Nigeria*, London, International African Institute.

BORCHERT, G. (1967) *Die Wirtschaftsräume Angolas*, Hamburg.

BORCHERDT, H. C. (1965) *Agrargeographie: Westermanns Lexikon der Geographie*, Braunschweig.

BORCHERDT, H. C. (1966) 'Zur Frage der Systematik landwirtschaftlicher Betriebsformen', *Berichte zur deutschen Landeskunde*, **36**, 95–100.

BORN, M. (1965) 'Zentralkordofan—Bauern und Nomaden in Savannengebieten des Sudans', *Marburger Geograph. Schriften*, **25**.

BRAIDWOOD, R. J. (1962) *Courses Toward Urban Life: archaeological considerations of some cultural alternates*, Edinburgh, University Press.

BRANDT, H., SCHUBERT, B. and GERKEN, E. T. (1972) *The Industrial Town as Factor of Economic and Social Development: the example of Jinja/Uganda*, Munich.

BRIGGS, G. W. G. (1941) 'Soil deterioration in the Southern District of Tiv Division, Benue Province', *Farm and Forest*, **2**.

BROMLEY, R. J. (1972) 'Agricultural colonisation in the Upper Amazon Basin', *Tijdschrift voor Econ. en Soc. Geografie*, **63**, 278–94.

BRONGER, D. (1970) 'Der sozialgeographische Einfluss des Kastenwesens auf Siedlung und Agrarstruktur im südlichen Indien', *Erdkunde*, **24**, 89–106; 194–207.

BROWN, L. R. (1970) *Seeds of Change: The Green Revolution and development in the 1970s*, New York, Praeger.

BRÜCHER, W. (1968) 'Die Erschließung des tropischen Regenwaldes am Ostrand der Kolumbianischen Anden', *Tübinger Geogr. Studien*, **28**, 218.

BRÜCHER, W. (1970) 'Rinderhaltung im Amazonischen Regenwald', *Tübinger Geogr. Stud.*, **34**, 215–27.

BUCHANAN, F. (1807) *A Journey from Madras through the Countries of Mysore, Canara and Malabar*, London.

BUCHANAN, K. M. (1964) 'Profiles of the third world', *Pacific Viewpoint*, **5**, 97–126.

BUDYKO, M. J. (1958) *The Heat Balance of the Earth's Surface*, US Dept. of Commerce, Weather Bureau, Washington.

BURNETT, J. R. (1948) 'Crop production', *Agriculture in the Sudan*, ed. J. D. Tothill, London, Oxford University Press.

BYAGAGAIRE, I. M. and LAWRANCE, I. C. D. (1957) *Land Tenure in Uganda*, Entebbe.

CANTOR, L. (1967) *A World Geography of Irrigation*, Edinburgh, Oliver & Boyd.

CAROL, H. (1952) 'Das agrargeographische Betrachtungssystem dargelegt am Beispiel der Karru in Südafrika', *Geogr. Helvetica*, **7**, 17–67.

CHAKRAVARTI, A. K. (1973) 'Green Revolution in India', *Ann. Ass. Am. Geogr.*, **63**, 319–30.

CHANG, J. H. (1968) 'The agricultural potential of the humid tropics', *Geogr. Rev.*, **58**, 333–61.

CHARTER, C. F. (1953) *Proceedings of the Cocoa Conference*, Maps, London.

CHISHOLM, M. (1962) *Rural Settlement and Land Use*, London, Hutchinson.

CLARK, C. and HASWELL, M. (1970) *The Economics of Subsistence Agriculture*, London, Macmillan.

CLARKE, J. I. (1965; 2nd edn. 1972) *Population Geography*, Oxford, Pergamon Press.

CLAYTON, E. S. (1964) *Agrarian Development in Peasant Economies: some lessons from Kenya*, Oxford, Pergamon Press.

COENE, R. DE (1956) 'Agricultural settlement schemes in the Belgian Congo', *Trop. Afric.*, 1–12.

CONKLIN, H. C. (1954) 'An ethnoecological approach to shifting cultivation', *Trans. New York Acad. Sci.*, **17**.

CONKLIN, H. C. (1957) *Hanunóo Agriculture, a Report on an Integral System of Shifting Cultivation in the Philippines*, FAO Forestry Development Paper 12, Rome. FAO.

CONKLIN, H. C. (1961–2) 'The study of shifting cultivation', *Current Anthropology*, 27–58.

COOK, E. (1931) *A Geography of Ceylon*, London, Macmillan.

COOK, E. K. (1953) *Ceylon: its geography, its resources and its people*, London, Macmillan.

COOK, O. F. (1921) 'Milpa agriculture', *Ann. Report Smithsonian Institute 1919*, Washington.

COURTENAY, P. P. (1965) *Plantation Agriculture*, New York, Praeger; London, Bell.

CREDNER, W. (1935) *Siam: das Land der Tai*, Stuttgart.

CROWTHER, F. (1948) 'A review of experimental work', in *Agriculture in the Sudan*, ed. J. D. Tothill, London, Oxford University Press, 439–592.

CURRY-LINDAHL, K. (1972) *Conservation for Survival: an ecological strategy*, New York, London, Gollancz.

DASMAN, R. F. (1972) *Planet in Peril?* Paris, Unesco, 1–136.

DAVEAU, S. and RIBEIRO, O. (1973) *La Zone intertropicale humide*, Paris.

DAVIES, D. H. (1971) *Zambia in Maps*, London, University of London Press.

DELVERT, J. (1961) *Le paysan cambodien*, Paris.

DICKINSON, J. C. (1972) 'High diversity alternatives in tropical landscape management', *International Geography* (papers submitted to Int. Geog. Congr., Montreal), Toronto, 713–14.

DICKSON, K. B. (1969) *A Historical Geography of Ghana*, London, Cambridge University Press.

DIETZEL, K. H. (1938) 'Grundfragen der Wirtschaftsorganisation in tropischen Kolonialländern', *Geogr. Zeitschr.*, **44**, 441–58.

DITTMER, K. (1965) 'Zur Entstehung des Rinderhirtennomadismus', *Paideuma*, **6**, 8–23.

DOBBY, E. H. G. (1950; 2nd edn., 1954) *South-East Asia*, London, University of London Press.

DOLLFUS, O. (1967) *Le Pérou: Introduction géographique á l'étude du développement*, Paris.

DRESCH, J. (1966) 'Utilization and human geography of the deserts', *Trans. Inst. British Geographers*, **40**, 1–10.

DUCKHAM, A. N. and MASEFIELD, G. B. (1971) *Farming Systems of the World*, London, Chatto & Windus.

DUMONT, R. (1957) *Types of Rural Economy; studies in World agriculture*, trans. D. Magnin, London, Methuen.

DUMONT, R. (1962) *L'Afrique Noire est mal partie*, Paris.

DUPUIS, J. (1960) *Madras et le Nord du Coromandel: étude des conditions de la vie indienne dans un cadre géographique*, Paris.

ECKAUS, R. S. (1955) The factor proportions problem in underdeveloped areas, *American Economic Review*.

EICHER, C. (1970) *Research on Agricultural Development in five English-Speaking Countries in West Africa*, Agricultural Development Council.

EKWALL, E. (1955) 'Slash-and-burn cultivation: a contribution to anthropological terminology', *Man*, 135–6.

ENGELBRECHT, H. (1899) *Die Landbauzonen der Erde*, Berlin.

ENGELBRECHT, H. (1930) 'Die Landbauzonen der Erde', *Pet. Mitt.*, complete issue, **209** (Herrmann Wagner Gedächtnisschrift), Gotha, 287–97.

ENGELHARD, K. and LIENAU, C. 'Der Zuckerrohranbau in Ostafrika', *Geogr. Rundschau*, **22**, 65–73.

ERZ, W. (1967) *Wildtierschutz und Wildtiernutzung in Rhodesien und im übrigen südlichen Afrika*, Munich, Ifo-Institut.

FAO *see* FOOD AND AGRICULTURE ORGANISATION.

FALKNER, F. (1938) 'Die Trockengrenze des Regenfeldbaus in Afrika', *Pet. Mitt.*, **84**, 209–14.

FARMER, B. H. (1950) 'Agriculture in Ceylon', *Geogr. Rev.*, **40**, 42–66.

FARMER, B. H. (1957) *Pioneer Peasant Colonization in Ceylon: a Study in Asian Agrarian Problems*, Oxford University Press for Royal Institute of International Affairs.

FARMER, B. H. (1963) 'Peasant and plantation in Ceylon', *Pacific Viewpoint*, **4**, 9–16.

FAUCHER, D. (1949) *Géographic Agraire: types de cultures*, Paris.

FAUTZ, B. (1970) 'Agrarräume in den Subtropen und Tropen Australiens', *Geogr. Rundschau*, **22**, 386–91.

FELS, E. (1965) 'Die Bewässerungsfläche der Erde', in *Festschrift Leopold G. Scheidl*, Vienna, 33–49.

FINCK, A. (1963) *Tropische Böden*, Hamburg, Berlin.

FISCHER, A. (1925) 'Zur Frage der Tragfähigkeit des Lebensraumes', *Zeitschr. f. Geopolitik*, **2**, 762–79.

FISHER, C. A. (1964) *South-East Asia: a social, economic and political geography*, London, Methuen.

FLOHN, H. (1957) 'Zur Frage der Einteilung der Klimazonen', *Erdkunde*, **11**, 161–75.

FLOHN, H. (1958) 'Rezension zu Budyko, 1958', *Erdkunde*, **12**. 233–6.

FLOYD, B. N. (1965) 'Terrace agriculture in Eastern Nigeria, a geographical appraisal', *Nigerian Geogr. Journ.*, **8**, 33–44.

FLOYD, B. (1970) 'Agricultural innovation in Jamaica: the Yallahs Valley Land Authority', *Econ. Geogr.*, **46**, 63–77.

FOOD AND AGRICULTURAL ORGANISATION (FAO) (1963) *Third World Food Survey*, Rome.

FOOD AND AGRICULTURE ORGANISATION (1966) *The State of Food and Agriculture*, Rome.

FOOD AND AGRICULTURE ORGANISATION (1957) *Shifting Cultivation in Urasylva*, Rome.

FOOD AND AGRICULTURE ORGANISATION (1962) *Africa Survey: Report on the possibilities of African rural development in relation to economic and social growth*, Rome.

FORTES, M., STEEL, R. W. and ADY, P. (1947) 'Ashanti Survey 1945–46; an experiment in social research', *Geogr. J.*, **110**, 149–79.

FOUND, W. C. (1971) *A Theoretical Approach to Rural Land Use Patterns*, London, Edward Arnold.

FOWLER, F. J. (1950) 'Some problems of water distribution between East and West Punjab', *Geogr. Rev.*, **40**, 583–99.

FRANCA, A. (1956) *Guide of Excursion 3: (the Coffee Trail and Pioneer Fringes)*, Internat. Geogr. Congr., Rio de Janeiro.

FREEMAN, J. D. (1955) *Iban Agriculture: a report on the shifting cultivation of hill rice by the Iban of Sarawak*, London, HMSO.

FREYRE, G. (1965) *Herrenhaus und Sklavenhütte*, Cologne.

FRITSCH, B. (1965) 'Die ökonomische Theorie als Instrument der Entwicklungspolitik', *Kyklos*, no. 18.

FRITSCH, B. (1968) *Entwicklungsländer*, Cologne/Berlin.

FRYER, D. W. (1958) 'World income and types of economies', *Econ. Geogr.*, **34**, 283–303.

FRYER, D. W. and JACKSON, J. C. (1966) 'Peasant producers or urban planters? The Chinese rubber smallholders of Ulu Selangor', *Pacific Viewpoint*, **7**, 198–228.

GALBRAITH, J. K. (1963) *Economic Development in Perspective*, London.

GALBRAITH, J. K. (1967) *The New Industrial State*, London, Hamish Hamilton Penguin, 1968.

GALLWITZ, K. (1963) Landtechnik in Afrikanischen Ländern 1961–2, *Afrika heute, Jahrbuch d. deutschen Afrikagesellschaft*, 135–60.

GARNIER, B. J. (1958) 'Some comments on defining the humid tropics', *Research Notes*, **2**, Ibadan, 9–25.

GARNIER, B. J. (1961a) 'The idea of humid tropicality', *Proceedings of the Tenth Pacific Science Congress, Honolulu*.

GARNIER, B. J. (1961b) 'Mapping the humid tropics: climate criteria', in 'Delimitation of the Humid Tropics', *Geogr. Rev.*, **51**, 339–46.

GARRISON, W. L. and MARBLE, D. F. (1957) 'The special structure of agricultural activities', *Ann. Ass. Am. Geogr.*, **47**, 137–44.

GERLING, W. (1954) *Die Plantage: Fragen ihrer Entstehung, Ausbreitung und wirtschaftlichen Eigenart*, Würzburg.

GHOSH, D. (1946) *Pressure of Population and Economic Efficiency in India*, New Delhi.

GILLMANN, C. (1936) 'A population map of Tanganyika Territory', *Geogr. Rev.*, **26**, 353–75.

GINSBURG, N. S. (1960) *Essays on Geography and Economic Development*, Chicago.

GINSBURG, N. S. (1961) *Atlas of Economic Development*, University of Chicago Press.

GLASER, G. (1971) 'Neue Aspekte der Rinderweidewirtschaft in Zentralbrasilien', *Beiträge zur Geographie Brasiliens*, Heidelberg, 19–38.

GLEAVE, M. B. (1963) 'Hill settlements and their abandonment in Western Yoruba Land', *Africa*, **30**, 4.

GLEAVE, M. B. (1966) 'Hill settlements and their abandonment in tropical Africa', *Trans. Inst. British Geographers*, **40**, 39–49.

GOULD, P. R. and SPARKS, J. P. (1969) 'The geographical context of human diets in southern Guatemala', *Geogr. Rev.*, **59**, 58–82.

GOURDON, P. R. T. (1914) *The Khasis*, London.

GOUROU, P. (1936) 'Les Paysans du Delta Tonkinois', in *Publ. de l'Ecole Franç. d'Extreme-Orient*, **26**; reprint, Paris, 1966.

GOUROU, P. (1947, 1953, 1966) *Les Pays Tropicaux*, Paris (1947); Engl. translation as *The Tropical World* 1953 and London, Longman.

GOROU, P. (1953) *L'Asie*, Paris.

GOUROU, P. (1956) 'The quality of land-use of tropical cultivators', in Thomas (1956).

GREEN, L. P. and FAIR, T. I. D. (1962) *Development in Africa*, Johannesburg.

GREGOR, H. F. (1970) *Geography of Agriculture: themes in research*, Englewood Cliffs, Prentice-Hall.

GREGOR, H. F. (1972) 'The quasi-plantation as a conceptual model', in *International Geography* (papers submitted to Int. Geog. Congr., Montreal), Toronto, 722.

GREGOR, H. F. (1972) 'Terminology in typology—the problems of "plantation",' in *International Geography* (papers submitted to Int. Geog. Congr., Montreal), Toronto, 713–14.

GRIGG, D. B. (1970) *The Harsh Lands: a study in agricultural development*, London, Macmillan.

GROTEWOLD, A. (1959) 'Von Thünen in retrospect', *Econ. Geogr.*, **35**, 346–55.

GUTERSOHN, H. (1940) 'Sao Paulo—Natur und Wirtschaft', *Vierteljahresschr. d. Naturforsch. Ges. Zürich*, **85**, 149–255.

GUTH, W. (1957) *Der Kapitalexport in unterentwickelte Länder*, Basel.

HABERLAND, E. (1963) *Völker Süd-Äthiopiens*, Stuttgart, vol. 2.

HABERLAND, E. (1964) 'Feldbau und Nutzpflanzen in Neu-Guinea', *Kolloquiums-Vortrag Geogr. Inst. Gießen.*

HAHN, E. (1896) *Die Haustiere und ihre Beziehung zur Wirtschaft des Menschen*, Leipzig.

HAHN, E. (1913) 'Die Hirtenvölker in Asien und Afrika', *Geogr. Zeitschr.*, **19**, 305–19, 369–82.

HAHN, H. (1963) 'Soziale Lage und Entwicklungsmöglichkeiten im Bereich klein-bäuerlicher Besitzstruktur in Afghanistan', *Deutscher Geographentag*, Heidelberg, 249–58.

HANCE, W. A. (1964) *The Geography of Modern Africa*, Columbia University Press.

HARRIS, D. R. (1971) 'The ecology of Swidden cultivation in the Upper Orinoco rain Forest, Venezuela', *Geogrl. Rev.*, **61**, 475–95.

HARRISON-CHURCH, R. J. (1963) 'Observations on large scale irrigation development in Africa', *Agric. Econ. Bull. for Africa*, no. 4, ECA-FAO Addis Ababa.

HARRISON-CHURCH, R. J. (1964) 'The Limpopo scheme', *Geogr. Mag.*, **37**, 212–27.

HARROY, J. P. (1949) *Afrique, terre qui meurt: la dégradation des sols africains sous l'influence de la colonisation*, Brussels.

HARVEY, D. W. (1966) 'Theoretical concepts and the analysis of agricultural land-use patterns in geography', *Ann. Ass. Am. Geogr.*, **56**, 361–74.

HECKLAU, H. (1968) 'Die agrarlandschaftlichen Auswirkungen der Boden-besitzreform in den ehemaligen White Highlands von Kenya', *Die Erde*, 236–64.

HENSHALL, J. D. (1967) 'Models of agricultural activity', in R. H. Chorley and P. Hagget, *Models in Geography*, London, Methuen.

HERZOG, R. (1963) 'Seßhaftwerden von Nomaden', *Forschungsbericht d. Landes Nordrhein-Westfalen*, Cologne.

HESMER, H. (1966) *Der kombinierte land- und forstwirtschaftliche Anbau, I. Trop. Afrika*, Stuttgart.

HESMER, H. (1970) *Der kombinierte land- und forstwirtschaftliche Anbau, II. Tropisches und subtropisches Asien*, Stuttgart.

HETTNER, A. (1929) *Der Gang der Kulturen auf der Erde*, Leipzig, Berlin.

HETTNER, A. (1930) 'Die Klimate der Erde', *Geogr. Schriften*, **5**, Leipzig, Berlin.

HETTNER, A. (1934) 'Die Klimate der Erde', *Vergleichende Länderkunde*, **3**, Leipzig, Berlin, 87–202.

HICKMANN, G. M. and DICKINS, W. H. G. (1960) *The Lands and Peoples of East-Africa*, London, Longmans.

HIGGINS, B. (1959; 2nd edn., 1968) *Economic Development*, New York, Norton; London, Constable.

HIGHSMITH, R. M. (1965) 'Irrigated lands of the world', *Geogr. Rev.*, **55**, 382–9.

HILL, P. (1955) 'The Mortgaging of Gold Coast Cocoa Farms, 1950–54', West African Inst. Social and Econ. Res. Conf., Ibadan.

HILL, P. (1963) *Migrant Cocoa Farmers of Southern Ghana*, Cambridge University Press.

HILL, P. (1970) *Studies in Rural Capitalism in West Africa*, Cambridge University Press.

HILLS, T. L. (1968) *The Ecology of the Forest-Savanna Boundary*, Montreal.

HIRTH, P. (1921) *Grundzüge einer Geographie der künstlichen Bewässerung*, Halle.

HO, R. (1962) 'Mixed farming and multiple cropping in Malaya', *Journal of Tropical Geography*, **16**, 1–17.

HODDER, B. W. (1959) *Man in Malaya*, London, University of London Press.

HODDER, B. W. (1968) *Economic Development in the Tropics*, London, Methuen.

HÖTZEL, D. (1963) 'Probleme der modernen Ernährung', *Ergebn. Landwirtsch. Forsch. a.d. Justus-Liebig-Universität*, **5**.

HÖVERMANN, J. (1958) 'Bauerntum und bäuerliche Siedlungen in Äthiopien', *Die Erde*, **89**, 1–20.

HOFMEISTER, B. (1961) 'Wesen und Erscheinungsformen der Transhumance', *Erdkunde*, **15**, 121–35.

HOLDRIDGE, L. R. (1959) 'Ecological indications of the need for a new approach to tropical land use', in A. Samper, ed., *Symposia Interamericana*, Turrialba, Costa Rica, 1–12.

HOLLSTEIN, W. (1937) 'Eine Bonitierung der Erde auf landwirtschaftlicher und bodenkundlicher Grundlage, *Pet. Mitt.*, Gotha, complete issue, **234**.

HOSELITZ, B. F. (1952) *The Progress of Underdeveloped Areas*, University of Chicago Press.

HUGHES, C. C. and HUNTER, J. M. (1972) 'Development and disease in Africa', *The Ecologist*, **2**, no. 10.

HUMBERT, H. (1927) 'La destruction d'une flore insulaire par le feu: principaux aspects de la végétation á Madagascar'. *Mémoires de l'Académie Matgäche*, Tananarive, **5**, 79.

HUNTER, J. M. (1963) 'Cocoa migration and patterns of land-ownership in the Densu Valley, Ghana', *Trans. Inst. Brit. Geogr.*, **33**, 6–87.

HUNTER, J. M. (1966) 'Ascertaining population carrying capacity under traditional systems of agriculture in developing countries', *The Professional Geographer*, **18**, 151–4.

HUNTER, J. M. (1973) 'Geophagy in Africa and in the United States', *Geogr. Rev.*, **63**, 170–95.

HUPPERTZ, J. (1951) 'Viehhaltung und Stallwirtschaft bei den einheimischen Agrarkulturen in Afrika und Asien', *Erdkunde*, **5**, 36–51.

HUTCHINSON, SIR J. (1966) 'Land and human populations', *Advancement of Science*, **23**, 1–14.

HUTTENLOCHER, F. (1957) 'Sozialgeographische Räume', in *Studium Generale*, **10**, 590.

IGBOZURIKE, M. U. (1971) 'Ecological balance in tropical agriculture', *Geogr. Rev.*, **61**, 519–24.

INNES, F. C. (1972) 'Some thoughts on the aftermath of the plantation system in the Caribbean', in *International Geography 1972; Papers submitted to the 22nd Int. Geogr. Cong.*, Montreal, 434–6.

INNIS, D. Q. (1972) 'The efficiency of tropical, small farm agricultural practices', in *International Geography Congr.* Toronto (papers submitted to Int. Geog. Congr., Montreal), 729–31.

INTERNATIONAL BANK (1954) *The Economic Development of Nigeria*, Washington.

INTERNATIONAL BANK (1955) *The Economic Development of Malaya*, Washington.

INTERNATIONAL BANK (1960) *The Economic Development of Venezuela*, Baltimore.

INTERNATIONAL BANK (1961) *The Economic Development of Tanganyika*, Baltimore.

INTERNATIONAL BANK (1962a) *The Economic Development of Ceylon*, Baltimore.

INTERNATIONAL BANK (1962b) *The Economic Development of Uganda*, Baltimore.

INTERNATIONAL BANK (1963) *The Economic Development of Kenya*, Baltimore.

IZARD, F. and IZARD, M. (1958) *Bouna: Monographie d'un village Pana de la Vallée du Souron (Haute Volta)*, Bordeaux.

JAEGER, F. (1934) 'Versuch einer Anthropogeographischen Gliederung der Erdoberfläche', *Pet. Mitt.*, **80**, 353–6.

JAEGER, F. (1943a) 'Neuer Versuch einer Anthropogeographischen Gliederung der Erdoberfläche', *Pet. Mitt.*, **89**, 313–23.

JAEGER, F. (1943b) 'Trocknet Afrika aus?', *Geogr. Zeitschr.* **49**, 1–19.

JAEGER, F. (1945) 'Zur Gliederung und Benennung des tropischen Graslandgürtels', *Verhandl. d. Naturforsch, Ges.* Basel, **56**, part 2, 509–20.

JAMES, P. E. (1953) 'Trends in Brasilian agricultural development', *Geogr. Rev.*, **43**, 301–28.

JÄTZOLD, R. (1965) 'Die Nachwirkungen des fehlgeschlagenen Erdnußprojekts in Ostafrika', *Erdkunde*, **19**, 210–33.

JÄTZOLD, R. (1970) 'Ein Beitrag zur Klassifikation des Agrarklimas der Tropen', *Tübinger Geogr. Studien*, 57–69.

JÄTZOLD, R. and BAUM, E. (1968) *The Kilombero Valley*, Munich, Ifo-Institute.

JOHNSTON, B. F. (1958) *The Staple Food Economies of Western Tropical Africa*, Stanford University Press.

JONES, C. F. (1928–30) 'Agricultural regions of South America', *Econ. Geogr.*, **4**, 1–30, 159–86, 267–94; **5**, 109–40, 277–307, 390–421; **6**, 1–36.

JONES, W. O. (1959) *Manioc in Africa*, Stanford University Press.

JOOSTEN, J. H. L. (1962) *Wirtschaftliche und agrarpolitische Aspekte tropischer Landbausysteme*, Vortrag Göttingen.

JUNGHANS, K. H. (1968) 'Beginnende Marktverflechtung landwirtschaftlicher Betriebe in Indien', *Jahrbuch des Südasien—Instituts der Universität Heidelberg 1967–8*, Wiesbaden, 144–67.

KARIEL, H. G. (1966) 'A proposed classification of diet', *Ann. Ass. Am. Geogr.*, **56**, 68–80.

KAYSER, B. (1969) *L'Agriculture et la société rurale des regions tropicales*, Paris, SEDES.

KELLMAN, M. C. (1969) 'Some environmental components of shifting cultivation in upland Mindanao', *J. Trop. Geogr.*, **28**, 40–56.

KELLOG, C. E. (1950) 'Tropical soils', in *Trans. 4th Int. Congr. Soil Science*, 1950, 266–76.

KENYATTA, J. (1938) *Facing Mount Kenya: the tribal life of the Kikuyu*, London, Secker & Warburg.

KINZL, H. (1942) 'Die anthropogeographische Bedeutung der Gletscher und die künstliche Flurbewässerung in den peruanischen Anden', *Sitz. ber. europ. Geographen Würzburg*, 353–80.

KISSELMANN, E. (1965) 'Stand und Formen der Mechanisierung der Landwirtschaft in den asiatischen Ländern', Stuttgart.

KNALL, B. (1962) 'Wirtschaftserschließung und Entwicklungsstufen', in *Weltwirtschaft. Archiv*, Bd **88**, 238.

KÖPPEN, W. (1931) *Die Klimate der Erde*, Berlin und Leipzig.

KÖPPEN, W. and GEIGER, R. (1936) *Handbuch der Klimatologie*, Berlin, Part C.

KOHLER, J. M. (1971) 'Activités Agricoles et Changements sociaux dans l'Ouest-Mossi (Haute-Volta)', in Mém. ORSTOM, Paris 1971, no. 46.

KOLB, A. (1940) 'Die Reislandschaft auf den Philippinen', *Pet. Mitt.*, **86**, 113–24.

KOLB, A. (1941) 'Tarokultur, Naßbau und künstliche Feldterrassen', in *Zeitschr. f. Erdkunde*, **9**, 750–3.

KOLB, A. (1942) *Die Philippinen*, Leipzig.

KOLB, A. (1942) 'Kolonialwirtschaftliche Strukturwandlungen in den pazifischen Tropen', in *Lebensraumfragen eropäischer Völker*, vol. 2, Leipzig, 501–31.

KOLB, A. (1963) *Ostasien: China, Japan, Korea: Geographie eines Kulturerdteils*, Heidelberg, 608.

KOLB, A. (1966) 'Geofaktoren, Landschaftsgürtel und Wirtschaftserdteile', in *Heidelberger Studien zur Kulturgeographie*, 29–36.

KOLLMANSPERGER, F. (1965) *Von Afrika nach Afrika*, Mainz.

KOSTROWICKI, J. and HELBURN, N. (1967) *Agricultural typology – principles and methods*, in *IGU Bull.*, **18**.

KREBS, N. (1939) *Vorderindien und Ceylon, eine Landeskunde*, Stuttgart.

KREBS, N. (1951) *Vergleichende Länderkunde*, Stuttgart.

KREEB, K. (1964) *Ökologische Grundlagen der Bewässerungskulturen in den Subtropen*, Stuttgart.

– KÜCHLER, A. W. (1961) 'Mapping the humid tropics: vegetation criteria', in *Geogr. Rev.*, **51**, 346–7.

KÜCHLER, J. (1968) 'Penang', *Gießener Geogr. Schriften*, Vol. 13, Gießen.

KUHNHOLTZ-LORDAT (1939) *La Terre incendiée: essai d'agronomie comparée*, Nimes.

KULARATNAM, K. (1955) 'Ceylon, Land, Volk, Wirtschaft', *Geogr. Rundschau*, **7**, 396–400.

KULS, W. (1958) *Beiträge zur Kulturgeographie der südäthiopischen Seenregion*, Frankfurt/M.

KULS, W. (1963) 'Bevölkerung, Siedlung und Landwirtschaft im Hochland von Godjam (Nordäthiopien)', in *Frankfurter Geogr.*, no. 39, 1–77.

KULS, W. (1968) 'Jüngere Wandlungen in den Enseteanbaugebieten Südäthiopiens', *Acta Geographica*, Helsinki, 185–99.

LASSERRE, G. (1961) *La Guadeloupe: étude géographique*, Bordeaux.

LAUER, W. (1952) 'Humide und aride Jahreszeiten in Afrika und Südamerika und ihre Beziehung zu den Vegetationsgürteln', *Bonner Geogr. Abhandl.*, **9**, 15–98.

LAUER, W. (1956) 'Vegetation, Landnutzung und Agrarpotential in El Salvador', *Schriften des Geogr. Instituts der Universität Kiel*, **16**, no. 1, 98.

LAUER, W. (1961) 'Wandlungen im Landschaftsbild des südchilenischen Seengebiets seit Ende der spanischen Kolonialzeit', *Schriften d. Geogr. Inst. Univ. Kiel*, **20**, 227–76.

LAUER, W. (1970) 'Naturgeschehen und Kulturlandschaft in den Tropen (Beispiel Zentralamerika)', *Tübinger Geogr. Studien*, 83–105.

LAVAUDEN, L. (1931) 'Le déboisement et la végétation de Madagascar', *Revue de la Botanique Appliquée et Agriculture Coloniale*, no. 122, 817–24.

LEAKEY, L. S. B. (1952) *Mau Mau and the Kikuyu*, London, Methuen.

LEHMANN, G. and JUSATZ, H. J. (1965) *Die Arbeitsfähigkeit des Menschen im tropischen Klima*, Cologne.

LESER, H. (1970) 'Wandlungen der bevölkerungs- und wirtschafts-geografischen Verhältnisse im Lichte der historischen und politischen Entwicklung im südl. Afrika', *Geogr. Zeitschr.*, **58**, 198–213.

LIENAU, C. and UHLIG, H. (1967) *Materialien zur Terminologie der Agrarlandschaft*, Giessen, vol. 1.

LOETSCH, F. (1958) 'Der Einfluss des Brandrodungsbaues auf das Gefüge des Tropenwaldes und die Wasserführung der Ströme untersucht am Beispiel Nordthailands', *Erdkunde*, **12**, 182–205.

LORENZ, D. (1961) 'Zur Typologie der Entwicklungsländer', *Jahrb. f. Sozial-wissensch.*, **12**, 354–80.

MAAS, W. (1963) *Hinterindien, Jll. Länderkunde*, Gütersloh, 1186–1276.

MCILROY, R. J. (1963) *An Introduction to Tropical Cash Crops*, Oxford University Press.

MCNEISH, R. S. (1964) 'Ancient Mesoamerican civilization', *Science*, **143**.

MANSHARD, W. (1955) 'Entwicklungspläne in der Gambia', *Erdkunde*, **9**, 220–4.

MANSHARD, W. (1961a) *Die geographischen Grundlagen der Wirtschaft Ghanas*, Wiesbaden.

MANSHARD, W. (1961b) 'Afrikanische Waldhufen und Waldstreifenfluren', *Die Erde*, **92**, 246–58.

MANSHARD, W. (1961c) 'Land-use patterns and agricultural migration in central Ghana', *Tijdschr. Econ. Soc. Geogr.*, **52**, 225–30.

MANSHARD, W. (1962) 'Agrarsoziale Entwicklungen im Kakaogürtel von Ghana', *Abhandl. deutscher Geographentag Köln*, Wiesbaden, 190–201.

MANSHARD, W. (1963a) 'Tropisches und südliches Afrika', *Gr. illustrierte Länderkunde*, Gütersloh, 233–526.

MANSHARD, W. (1963b) 'Die Bedeutung der Geographie für Entwicklungsarbeiten in Tropisch-Westafrika', *Die Erde*, **94**, 225–46.

MANSHARD, W. (1965a) 'Landbesitz in Tropisch-Afrika – ein Beitrag zur geogr. Analyse der Agrarverfassung', *Giessener Geogr. Schriften*, **6**.

MANSHARD, W. (1965b) 'Kigezi—die agrargeographische Struktur eines ostafrikanischen Berglandes', *Erdkunde*, **19**, 192–210.

MANSHARD, W. (1965c) 'Die Viehhaltung in den Trockengebieten Tropisch-Afrikas', in *Weidewirtschaft in Trockengebieten, Gießener Beiträge zur Entwicklungsforschung—Schriftenreihe des Tropeninst. d. Justus-Liebig-Univ. Gießen*, vol. 1, 29–36.

MANSHARD, W. (1973) *Unsere gefährdete Umwelt*, Paderborn, 1–32.

MARBY, H. (1971) 'Die Teelandschaft der Insel Ceylon', in Erdkundl. Wissen, Wiesbaden, 23–101.

MARCH, G. F. (1936) 'The development of native agriculture in the Nuba mountain area of Kordofan Province, Anglo-Egyptian-Sudan', *Empire Journal of Experimental Agriculture*, Oxford.

MARSHALL, L. (1960) 'Kung Bushman Bands', in *Afrika*, London, 325–54.

MATZNETTER, J. (1963) 'Die Guineainsel Sao Thomé und Principe und ihre Plantagen', *Geogr. Zeitschr.*, **51**, 268–301.

MAULL, O. (1936) 'Die Bestimmung der Tropen am Beispiel Amerikas', in *Festschr. z. Hundertjahrfeier d. Vereins f. Geogr. u.Statistik zu Frankfurt/M*, 337–66.

MAY, J. M. (1952) 'Map of the world distribution of dengue and Yellow Fever', *Geogr. Rev.*, **42**, 282–23.

MAY, J. M. (1953) 'The mapping of human starvation, diets and diseases', *Geogr. Rev.*, **43**, 253–5, 403–4.

MECKELEIN, W. (1959) *Forschungen in der zentralen Sahara*, Braunschweig.

MECKELEIN, W. (1966) 'Entwicklungstendenzen der Kulturlandschaft im Industriezeitalter', in *Staatsanzeiger für Baden-Württemberg*, Stuttgart.

MEEK, C. K. (1957) *Land Tenure and Administration in Nigeria and the Cameroons*, London, HMSO.

MEGITT, M. J. (1963) 'Aboriginal food-gatherers in tropical Australia', IUCN

(Internat. Union Conserv. Nature) 9th Techn. Meeting, Nairobi.

MENSCHING, H. (1953) 'Formen der Eingeborenenwirtschaft in Marokko', *Die Erde*, **5**, 30–44.

MILNE, G. (1935) 'Some suggested units of classification and mapping, particularly of East-African soils', *Soil Research*, 183–98.

MIRACLE, M. P. (1967) *Agriculture in the Congo Basin*, University of Wisconsin Press.

MOHR, E. C. J. and BAREN, F. A. VAN (1954) *Tropical Soils*, The Hague.

MOMSEN, J. H. (1972) 'A model of agricultural change in developing areas', *International Geography* (papers submitted to Int. Geog. Congr., Montreal), *Congr. Toronto*, 1101–3.

MONHEIM, F. (1965) *Junge Indianerkolonisation in den Tiefländern Ostboliviens*, Braunschweig.

MONHEIM, F. (1966) 'Studien zur Haziendawirtschaft des Titicacabeckens', in *Heidelberger Studien zur Kulturgeographie (Gottfried-Pfeifer-Festschrift)*, 133–63.

MORAL, P. (1961) *Le Paysan Haïtien*, Paris.

MORGAN, W. B. and MUNTON, R. J. C. (1971) *Agricultural Geography*, London, Methuen.

MORGAN, W. B. and MOSS, R. P. (1968) 'Savanna and forest in Western Nigeria', *in Africa*, 286–93.

MORGAN, W. T. W. (1963) 'The "White Highlands" of Kenya', *Geogr. J.*, London, **129**, 140–55.

MORTIMORE, M. J. (1967) 'Land and population pressure in the Kano close-settled zone, northern Nigeria', *Adv. of Sci.*, **23**, 677–86.

MOSS, R. P. (1963) 'Soils, slopes and land use in a part of south-western Nigeria', *Trans. Inst. of British Geographers*, **32**, 143–68.

MÜLLER, J. O. (1967) 'Probleme der Auftrags-Rinderhaltung der Fulbe-Hirten (Peul) in West-Afrika', *Afrika-Studien*, Berlin, Heidelberg, **14**, 120.

MUKWAYA, A. B. (1953) 'Land tenure in Buganda', *East African Studies*, no. 1, Kampala.

MULLICK, M. A. (1972) 'Die Grüne Revolution in Pakistan', *Geogr. Rundschau*, **24**, 332–7.

MULLICK, M. A. (1973) 'Wie steht es um die Grüne Revolution?', *Entwicklung und Zusammenarbeit*, Bonn, **4**, 10–12.

MYINT, H. (1964) *The Economics of the Developing Countries*, London, Hutchinson.

MYRDAL, G. (1957) *Economy Theory and Underdeveloped Regions*, London, Duckworth (paperback, Methuen).

MYRDAL, G. (1957) *Rich Lands and Poor*, New York, Harper & Row.

NG KAY FONG, TAN CHEE LIAN, WIKKRAMATILEKE, R. (1966) 'Three farmers of Singapore', *Pacific Viewpoint*, **7**, 169–97.

NICOLAI, H. (1961) 'Luozi, etude géographique d'un pays du Bas-Congo', *Acad. Roy. Sci. d'Outre-Mer*.

NITZ, H. J. (1966) 'Formen bäuerlicher Landnutzung und ihre räumliche Ordnung im Vorderen Himalaya von Kumaon (Nord-West-Indien)', *Heidelberger Studien zur Kulturgeographie* Wiesbaden, 311–30.

NITZ, H. J. (1971) 'Formen der Landwirtschaft und ihre räumliche Ordnung in der oberen Gangesebene', *Heidelberger Geogr. Arbeiten*, **28**.

NYE, P. H. and GREENLAND, D. (1960) 'The soils under shifting cultivation', *Commonw. Agric. Bureau Techn. Comm.*, no. 51.

– OCHSE, J. J., SOULE, M. J., DIJKMAU, M. J. and WEHLBURG, C. (1961) *Tropical and Subtropical Agriculture*, New York, Collier-Macmillan, vol. 2.

O'CONNOR, A. M. (1966) *An Economic Geography of East Africa*, London, Bell.

OOI JIN-BEE (1959), 'Rural development in tropical areas with special reference to Malaya', *J. Trop. Geog.*, **12**, 1–222.

ORACION, T. S. (1963) 'Kaingin Agriculture among the Bukidnons of SE Negros, Philippines', *J. Trop. Geogr.*, **17**, 213–24.

OTREMBA, E. (1950) 'Die wirtschaftsgeographische Ordnung der Länder', *Die Erde*, **1**, 216–32.

OTREMBA, E. (1952) 'Grundbegriffe für die landwirtschaftsgeographische Arbeit in Mitteleuropa', *Geogr. Taschenb.*, 374–84.

OTREMBA, E. (1958) 'Die landwirtschaftlichen Betriebsformen in Venezuela und das Problem der Agrarkolonisation durch Europäer', *Wiss. Veröffentlichungen d. deutschen Instituts f. Länderkunde*, Leipzig, no.15/16, 5–50.

OTREMBA, E. (1960) *Allgemeine Agrar- und Industriegeographie*, 2nd edn., Stuttgart.

OTREMBA, E. (1962) 'Die raumwirtschaftliche Problematik der Entwicklungsländer', *Studium Generale*, 519–29.

OTREMBA, E. (1969) *Der Wirtschaftsraum—seine geographischen Grundlagen und Probleme*, Stuttgart, Franckh'sche Verlagshandlung.

OTREMBA, E. and KESSLER, M. (1965) Die Stellung der Viehwirtschaft im Agrarraum der Erde, Wiesbaden.

PAFFEN, K. H. and TROLL, C. (1963, 1965) *Weltkarten zur Klimakunde*, Berlin, Göttingen, Heidelberg.

PARSONS, J. J. and BOWEN, W. A. (1966) 'Ancient ridged fields of the San Jorge river floodplain, Colombia', *Geogr. Rev.*, **56**, 317–43.

PEDRAZA, G. J. W. (1956) 'Land consolidation in the Kikuyu areas of Kenya', *J. Afr. Adm.*, **8**, no. 2.

PELISSIER, P. (1966) *Les Paysans du Sénégal. Les civilisations agraires du Cayor à la Casamance.* Saint-Yriex la Perche.

PELZER, K. J. (1935) *Die Arbeiterwanderungen in Südostasien*, Hamburg, Friederichsen, de Gruyter & Co.

PELZER, K. J. (1948) *Pioneer Settlement in the Asiatic Tropics*, Amer. Geog. Soc. Special Publ. no. 29, New York.

⤳ PELZER, K. J. (1957) 'Agriculture in the humid tropics', *Proceedings of the Ninth Pacific Science Congress*, Bangkok, 20, 124–43.

PENCK, A. (1924) 'Das Hauptproblem der physischen Anthropogeographie', *Sitzungsberatung d. Preußischen Akademie d. Wissenschaften*, Berlin.

PENCK, A. (1941) 'Die Tragfähigkeit der Erde', in K. H. Dietzel (ed.), *Lebensraumfragen Europäischer Völker*, Leipzig, vol. 7, 10–32.

PETERS, D. V. (1950) *Land Usage in Serenje District*, in Rhodes-Livingstone Papers no. 19, London.

PFEIFER, G. (1928) 'Über raumwirtschaftliche Begriffe und Vorstellungen und ihre bisherige Anwendung in der Geographie und Wirtschaftswissenschaft', *Geogr. Zeitschr.*, **34**, 321–40; 411–25.

PFEIFER, G. (1939) 'Sinaloa und Sonora, Beitrag zur Landeskunde und Kulturgeographie des nordwestlichen Mexico', *Mitt. Geogr. Ges. Hamburg*, **46**.

PFEIFER, G. (1948) 'Die Ernährungswirtschaft der Erde', *Abhandlungen d. deutschen Geographentags*, Munich. 241–70.

PFEIFER, G. (1952) 'Das Wirtschaftsgeographische Lebenswerk Leo Waibels, geb. 22.2.1888, gest. 4.9.1951', *Erdkunde*, **6**, 1–20.

PFEIFER, G. (1953) 'Landwirtschaftliche Betriebssysteme und Kolonisationserfolg in Südbrasilien auf Grund der Forschungen von Leo Waibel', *Erdkunde*, **7**, 241–9.

PFEIFER, G. (1956) 'The quality of peasant living in central Europe', in Thomas (1956), 240–77.

PFEIFER, G. (1958) 'Zur Funktion des Landschaftsbegriffes in der deutschen Landwirtschaftsgeographie', *Studium Generale*, 399–411.

PFEIFER, G. (1962) 'Brasilien als Entwicklungsland—Beobachtungen im Hinterland von Rio, in Espirito Santo, Minas Gerais, Goias und Amazonien', *Entwicklungshilfe und Entwicklungsland—Westfälische Geogr. Studien*, Münster, **15**, 125–94.

PHILIPPSON, A. (1933) *Grundzüge der allgemeinen Geographie*, Leipzig, vol. 1.

PHILLIPS, J. (1959) *Agriculture and Ecology in Africa*, London, Faber.

PHILLIPS, J. (1961; 2nd edn. 1966) *The Development of Agriculture and Forestry in the Tropics*, London, Faber.

PLATT, R. S. (1935) 'Coffee plantations of Brazil: a comparison of occupance patterns in established and frontier areas', *Geogr. Rev.*, **25**, 231–9.

POPENOE, H. (1957) 'The influence of shifting cultivation cycle on soil properties in Central America', *Ninth Pac. Sci. Congr.*, Bangkok, 72–7.

POPENOE, H. (1963) *The Pre-Industrial Cultivator in the Tropics*, IUCN (*Internat. Union Conserv. Nature*), Nairobi, pamphlet.

PRESTON, E. J. (1953) 'Trends in Brasilian agricultural development', *Geogr. Rev.*, **43**, 301–28.

PRESTON, N. A. (1965) 'Negro, mestizo and Indian in an Andean environment', *Geogr. J.*, **131**, 220–65.

PROTHERO, R. M. (1957) 'Land-use at Soba, Zaria Province, Northern Nigeria', *Econ. Geogr.*, **33**, 72–86.

PROTHERO, R. M. (1958) *Migrant Labour from Sokoto Province, Northern Nigeria*, Kaduna.

PROTHERO, R. M. (1969) *A Geography of Africa*, London, Routledge.

RAHMANN, M. (1967) 'Probleme der Be- und Entwässerung, Versalzung und Vernässung im Sind (West-Pakistan)', *Geogr. Rundschau*, **19**, 261-5.

RAHMAN, M. U. (1972) 'Irrigation agriculture in Sind, Pakistan', International Geography (papers submitted to Int. Geog. Congr., Montreal), Toronto, 778–80.

RAULIN, H. (1967) 'La dynamique des techniques agraires en Afrique tropicale du Nord', CNRS, Paris, 202.

RAULIN, H. (1969) 'Communautés d'entraide et développement agricole au Niger—l'exemple de la Majya', *Etudes Rurales*, 5–26.

RICHARD-MOLARD, J. (1952) 'Groupements éthniques et civilisations nègres d'Afrique', *Cahiers d'Outre Mer*, **5**, 5–25.

RICHARDS, A. I. (1939) *Land, Labour and Diet in Northern Rhodesia; an economic study of the Bemba tribe*, London, Oxford University Press.

RICHARDS, A. I. (1954) *Economic Development and Tribal Change: a study of immigrant labour in Buganda*, Cambridge, Heffer.

RICHARDS, A. I. (1958) 'A changing pattern of agriculture in East Africa: the Bemba of Northern Rhodesia', *Geogr. Journ.*, **124**, 302–14.

RICHTHOFEN, F. VON (1908) *Vorlesungen über allgemeine Siedlungs- und Verkehrsgeographie*, Berlin.

RINGER, K. E. (1963) *Agrarverfassung im tropischen Afrika*, Freiburg.

RITTER, C. (1822) *Die Erdkunde im Verhältnis zur Natur und zur Geschichte des Menschen*, Berlin.

RODENWALDT, E. (1952–61) *Welt-Seuchen-Atlas*, Hamburg.

ROLFES, M. and WOERMANN, E. (1954) 'Landwirtschaftliche Betriebssysteme', *Handbuch der Landwirtschaft*, **5**, 345–62.

ROSTOW, W. W. (1953; 2nd edn., 1960) *The Process of Economic Growth*, Oxford University Press.

ROSTOW, W. W. (1960, 1971) *The Stages of Economic Growth*, Cambridge University Press.

ROSTOW, W. W. (1956) 'The take-off into self-sustained growth', *Economic Journal*.

ROTENHAN, D. FREIHERR VON. (1966) 'Bodennutzung und Viehhaltung im Sukuma-Land (Tansania)', *Afrika-Studien*, Berlin, Heidelberg, **11**, 131.

ROUGERIE, E. A. G. (1963) 'Das Problem des Bauerntums im Waldgebiet der Elfenbeinküste', *Die Erde*, **94**, 265–80.

ROUNCE, N. V. (1949) *The Agriculture of the Cultivations Steppe of the Lake, Western and Central Province of Tanganyika*, Kapstadt.

RÜHL, A. (1925) *Vom Wirtschaftsgeist im Orient*, Leipzig.

RUTHENBERG, H. (1963) 'Produktion unter genauer Aufsicht', in *Schmollers Jahrbuch*.

RUTHENBERG, H. (1965) 'Problem des Übergangs vom Wanderfeldbau und semipermanenten Feldbau zum permanenten Trockenfeldbau in Afrika südlich der Sahara', *Agrarwirtschaft*, Hannover, 25–32.

RUTHENBERG, H. (1967) 'Organisationsformen der Bodennutzung und Viehhaltung in den Tropen und Subtropen, dargestellt an ausgewählten Beispielen, Stutgart', 122–208. In Blanckenburg (Ed) 1967.

RUTHENBERG, H. (1968) *Smallholder Farming and Smallholder Development in Tanzania*, Munich, Ifo-Institute.

RUTHENBERG, H. (1971) *Farming Systems in the Tropics*, Oxford, Clarendon Press.

SALIN, E. (1959) 'Unterentwickelte Länder-Begriff und Wirklichkeit', *Kyklos*, 402–24.

SANDNER, G. (1961) 'Die ungelenkte bäuerliche Urwaldkolonisation in Costa Rica', *Abhandlung d. deutschen Geographentags Köln*, 202–10.

SANDNER, G. (1961a) 'Agrarkolonisation in Costa Rica', *Schriften d. Geogr. Inst. d. Universität Kiel*, **19**, 1–199.

SANDNER, G. (1961b) 'Das Valle General—landeskundliche Skizzen eines jungen Rodungsgebietes in Costa Rica', *Schriften d. Geogr. Instituts d. Universität Kiel*, **20**, 125–65.

SANDER, G. (1962) 'Costa Rica: Entwicklung, Struktur und Probleme seiner Wirtschaft', *Geogr. Taschenb.*, 205–24.

SANDNER, G. (1964) 'Die Erschließung der karibischen Waldregion im südlichen Zentralamerika', *Die Erde*, **95**, 111–31.

SANDNER, G. (1966) 'Aufbau, Arbeitsmethoden und Aufgaben der Zentralstelle für Angewandte Geographie am Instituto de Tierras y Colonización in Costa

Rica', *Nürnberger Wirtschafts-und Sozialgeographischen Arbeiten*, **5**, 65–76.

SANDER, G. (1970) 'Ursachen und Konsequenzen wachsenden Bevölkerungs-drucks im zentralamerikanischen Agrarraum', *Tübinger Geographische Studien* 1970, 279–92.

SAPPER, K. (1923) *Die Tropen*, Stuttgart.

SAPPER, K. (1932) 'Die Verbreitung der künstlichen Feldbewässerung', *Pet. Mitt.*, **78**, 225–31; 295–301.

SAPPER, K. (1934) 'Geographie der altindianischen Landwirtschaft', *Pet. Mitt.*, **80**, 41–5.

SAUER, C. O. (1952) *Agricultural Origins and Dispersals*, Am. Geogr. Soc., Bowman Mem. Lecture 2.

SAUTTER, G. (1951) 'Une économie indigène progressive—les Bacongo du District de Boko (Moyen-Congo)', *Bull. Assoc. des Géogr. Franç.*, no. 216–7, 64–72.

SAUTTER, G. (1951) 'Terroirs tropicaux', in *Structures agraires et paysages ruraux*, Mem. no. 17 des Annales de l'Est, Nancy, 119–61.

SAUTTER, G. (1962) 'A propos de quelques terroirs d'Afrique Occidentale', *Etudes Rurales*, 24–86.

SAUTTER, G. (1966) *De l'Atlantique au fleuve Congo: Une géographie du souspeuplement*, Paris.

SCHEIDL, L. (1963) 'Die Probleme der Entwicklungsländer in wirtschaftsgeographi-scher Sicht', *Wiener geograph. Schriften*, **16**, 5–49.

SCHILLER, O. (1964) 'Agrarstruktur und Agrarreform in den Ländern Süd- und Südostasiens, Hamburg, Berlin 1964, 128.

SCHLIPPE, P. DE (1956) *Shifting Cultivation in Afrika*, London, Routledge.

SCHMIEDER, O. (1962) *Die Neue Welt—1. Teil: Mittel- und Südamerika*, Heidelberg, Munich.

SCHMITTHENNER, H. (1951) *Lebensräume im Kampf der Kulturen*, Heidelberg.

SCHWEINFURTH, U. (1966) 'Die Teelandschaft im Hochland der Insel Ceylon als Beispiel für den Landschaftswechsel', in *Heidelberger Studien zur Kulturgeographie* Wiesbaden, 297–310.

SCHWEINFURTH, U. (1969) 'Pyrethrum cultivation: an attempt at development in the Central Cordillera of Eastern New Guinea', in *Yearbook of the South Asia Institute, Heidelberg University*, 1968/69.

SCHWEINFURTH, U., MARBY, H., WEITZEL, K., HAUSHERR, K. and DOMROS, M. (1971) *Landschaftsökologische Forschungen auf Ceylon*, Wiesbaden.

SHANTZ, H. L. (1940–3) 'Agricultural regions of Africa', *Econ. Geogr.*, **16**, 1–47, 122–61, 341–89; **17**, 217–49, 353–79; **18**, 229–46, 343–62; **19**, 77–109, 217–69.

SHANTZ, H. L. (1956) 'History and problems of arid lands development', in *The Future of Arid Lands*, Am. Soc. Adv. Sci., Washington.

SICK, W. D. (1959) 'Beiträge zur wirtschaftsräumlichen Gliederung Ecuadors', *Abhandlung Deutscher Geographentag*, Berlin, 270–6.

SICK, W. D. (1961) 'Ecuador: wirtschaftlicher Strukturbericht', *Geogr. Taschenb.*, 1960/61.

SICK, W. D. (1963) *Wirtschaftsgeographie von Ecuador*, Stuttgart.

SIEVERS, A. (1964) *Ceylon—Gesellschaft und Lebensraum in den orientalischen Tropen*, Wiesbaden.

SIOLI, H. (1969) 'Entwicklung und Aussichten der Landwirtschaft im brasilianischen Amazonasgebiet', *Die Erde*, **100**, 307–26.

SMITH, C. (1973) 'A case study of shifting cultivation and regional development in northern Tanzania', in *Zeitschrift für ausländische Landwirtschaft* Frankfurt/M, 22–39.

SOPHER, D. E. (1964) 'The Swidden wet rice transition zone in the Chittagong Hills', *Ann. Ass. Am. Geogr.*, **54**, 107–26.

SORRE, M. (1948) *Les Fondements de la géographie humaine*, vols. 1, 2; *Les Techniques de la vie sociale*, Paris.

SORRE, M. (1952) *Les fondements de la géographie humaine*, vol. 3; L'Habitat, Paris.

SORRE, M. (1952) 'La géographie de l'alimentation', *Annales de Géographie*, **61**, 184–99.

SORRENSON, M. P. K. (1963) 'Counter revolution to Mau Mau: land consolidation in Kikuyuland 1952–60', Proc. East Afr. Inst. Soc. Res. Conf., Kampala 1963.

SPATE, O. H. K. (1954) *India and Pakistan*, London, Methuen.

SPENCER, J. E. (1966) *Shifting Cultivation in Southeastern Asia*, Univ. of California Publ. in Geography, no. 19.

STEBBING, E. P. (1922) *The Forests of India*, London, Lane.

STEBBING, E. P. (1935) 'The encroaching Sahara—the threat to the West-African colonies', *Geogr. J.*, **85**, 506–24.

STEEL, R. W. (1966) 'Geography and the developing world', *Advanc. of Science*, Nottingham, **23**, 1–11.

STOUSE, P. A. D. (1970) 'Instability of tropical agriculture: the Atlantic lowlands of Costa Rica', *Econ. Geogr.*, **60**, 78–97.

SUPAN, A. (1879) 'Die Temperaturzonen der Erde', *Pet. Mitt.*, **13**.

SUPAN, A. (1922) *Leitlinien der allgemeinen politischen Geographie*, Berlin, Leipzig.

SUTER, K. (1952) 'Zur Anthropogeographie einer Oase des algerischen Sahara', *Mitt. Österr. Geog. Ges.*, Vienna, **94**, 31–54.

SUTER, K. (1954) 'Der Gartenbau in Touat', *Pet. Mitt.*, **98**, 176–84.

TANAKA, K. (1957) 'Japanese immigrants in Amazonia and their future', *Kobe Univ. Econ. Rev.*, **3**, 1–23.

TEMPANY, H. and GRIST, D. H. (1958) *An Introduction to Tropical Agriculture*, London, Longmans.

TERRA, G. J. A. (1953) 'Some sociological aspects of agriculture in S-E Asia', *Indonesië*, **6**, 439–63.

TERRA, G. J. A. (1953) 'The distribution of mixed gardening in Java', *Landbouw*, **25**.

TERRA, G. J. A. (1956) 'Tropische Landbouw en welvaart', *Economie*, **20**, 404–39.

TERRA, G. J. A. (1957) 'Landbouwstelsels en bedrijstelsels in de tropen', *Landbouw, Tijdschr.*

TERRA, G. J. A. (1958) 'Farm system in South-East Asia', *Netherl. J. Agric. Sci*, **6**.

TERRA, G. J. A. (1959) 'Agriculture in economically underdeveloped regions, especially in equatorial and subtropical regions', *Netherl. Journ. Agric. Sci.*, **7**, 216–31.

THOMAS, L. T. (1956) in *Man's Role in Changing the Face of the Earth* (International Symposium), University of Chicago Press.

THORNTHWAITE, C. W. (1948) 'An approach toward a rational classification of climate', *Geogr. Rev.*, **38**, 55–94.

THÜNEN, J. H. VON (1826) *Der isolierte Staat in Beziehung auf Landwirtschaft und Nationalökonomie*, Hamburg; reprinted Stuttgart, 1966.

TIMMERMANN, O. F. (1935) 'Ceylon', *Münchner Geogr. Ges.*, **28**, 169–323.

TONDEUR, G. (1955) 'Shifting cultivation in the Belgian Congo', *Unasylva*, Rome.

TOSI, J. A. and VOERTMAN, R. F. (1964) 'Some environmental factors in the economic development of the tropics', *Econ. Geogr.*, **40**, 189–205.

TOTHILL, J. D. (1940, 1956) *Agriculture in Uganda*, London, Oxford University Press; FAO, Rome, 1956.

TOYNBEE, A. J. (1934–9) *A Study of History*, Oxford University Press.

TRAPNELL, C. G. (1953) *The Soils, Vegetation and Agriculture in North-Eastern Rhodesia*, Lusaka.

TREWARTHA, G. T. (1943) *An Introduction to Weather and Climate*, 2nd edn., New York, McGraw-Hill.

TRICART, J. (1959) 'Les exchanges entre la zone forestière de côte d'Ivoire et les savanes soudaniennes', *Cahiers d'Outre Mer.*, **9**, 209–38.

TROLL, C. (1930) Die wirtschaftsgeographische Struktur des tropischen Süd-amerika, *Geogr. Zeitschr.*, **36**, 468–85.

TROLL, C. (1943) 'Thermische Klimatypen der Erde', *Pet. Mitt.*, **89**, 81–9.

TROLL, C. (1950) 'Die geographische Landschaft und ihre Erforschung', *Studium Generale* 1950, 163–81.

TROLL, C. (1951) 'Das Pflanzenkleid der Tropen in seiner Abhängigkeit von Klima, Boden und Mensch', in *Abhandlung Deutscher Geographentag*, Frankfurt, 35–66.

TROLL, C. (1959) 'Die tropischen Gebirge—ihre dreidimensionale klimatische und pflanzengeographische Zonierung', *Bonner Geogr. Abhandlungen*, **25**, 93.

TROLL, C. (1963a) 'Landscape, ecology and land development with special reference to the tropics', *J. Trop. Geogr.*, **17**, 1–11.

TROLL, C. (1963b) 'Qanat-Bewässerung in der Alten und Neuen Welt', *Mitt Österreich Geogr. Ges.*, **105**, 313–30.

TROLL, C. (1964) 'Die soziale Landschaft der Entwicklungsländer', *Frankfurter Allgemeine Zeitung*, 31 October.

TROLL, C. (1966) 'Die räumliche Differenzierung der Entwicklungsländer in ihrer Bedeutung für die Entwicklungshilfe', *Erdkunde Wissen*, Wiesbaden, **13**.

TULIPPE, O. (1955) 'Les paysannats indigènes au Kasai', *Soc. Belge d'Et. géogr.*, **24**, 21–67.

TULIPPE, O. (1956) 'Une révolution agraire—les paysannats indigènes au Congo Belge', *Comptes Rendues*. 18th Int. Geog. Congr., Rio de Janeiro, vol. 4, 173–8.

TULIPPE, O. (1957) in *Essai de Géographie agraire à Matafu: premier rapport de la mission scientifique de l'Université de Liege au Katanga*, Katanga, 48–51.

TULIPPE, O. and WILMET, J. (1964) 'Géographie de l'agriculture en Afrique Centrale', *Bull. Soc. Belge d'Etudes Géograph.*, 303–74.

UDO, R. K. (1971) 'Food—deficit areas of Nigeria', *Geogr. Rev.*, **61**, 415–30.

UHLIG, H. (1962) 'Indien—Probleme und geographische Differenzierung eines Entwicklungslandes', *Giessener Geogr. Schr.*, **2**, 7–46.

UHLIG, H. (1963) 'Die Volksgruppen und ihre Gesellschafts- und Wirtschaftsent-wicklung als Gestalter der Kulturlandschaft in Malaya', *Mitt. Österreich. Geogr. Ges.*, **105**, 65–94.

UHLIG, H. (1965) 'Die geographischen Grundlagen der Weidewirtschaft in den Trockengebieten der Tropen und Subtropen', *Schriften d. Tropeninstituts d. Justus-Liebig*, Universität Giessen, **1**, 1–28.

UHLIG, H. (1966) 'Bevölkerungsgruppen und Kulturlandschaften in Nord-Borneo', in *Festschrift G. Pfeifer*, Wiesbaden, 265–96.

UHLIG, H. (1969) 'Hill tribes and rice farmers in the Himalayas and South East Asia', *Trans. Inst. Brit. Geogr.*, **47**, 1–23.

UHLIG, H. (1970) 'Die Agrarlandschaften des Chenab—Tals in Jammu und Kaschmir', *Tübinger Geogr. Studien*, 309–23.

– UNESCO (1958) *Problems of Humid Tropical Regions*, Paris.

– UNESCO (1958) *Study of Tropical Vegetation*, Paris.

UNESCO (1963) *Enquête sur les Ressources Naturelles du Continent Africain*, Paris.

UNITED NATIONS (1955) *Determinants and Consequences of Population Trends*, New York.

UNITED NATIONS (1958) *The Future Growth of World Population*, Population Studies no. 28, UN. Dept. of Econ. and Soc. Affairs, New York.

UNITED NATIONS (1960) *Demographic Yearbook*, New York.

UNITED NATIONS (1963) *Fertiliser Use: spearhead of agricultural development*, New York.

UNITED STATES DEPARTMENT OF AGRICULTURE (1970) *World Demand Prospects for Agricultural Exports of less Developed Countries, in 1980*, Washington.

VAN VALKENBURG, S. and HELD, C. C. (1952) *Europe*, 2nd edn., New York; London, Chapman & Hall.

VALVERDE, O. and DIAS, C. V. (1967) 'A Rodovia Belém—Brasilia—*Estudo de Geografia Regional*, Rio de Janeiro.

VALVERDE, O. (1968) 'A Amazônia Brasileira'. Finisterra, *Revista Portuguesa de Geografia*, Lisbon, **3**, 240–56.

– VALVERDE, O. (1971) 'Shifting cultivation in Brazil: ideas on a new land policy', *Heidelberger Geogr. Arbeiten*, **34**, 1–17.

VANDESYST, H. (1924) 'L'évolution des formations botanico-agronomiques dans le Congo Occidental', *Revue des questions scientifiques*, 65–83.

VERSTAPPEN, H. T. (1966) 'Land forms, water and land use west of the Indus plain', *Nature and Resources*, **3**.

VISHER, S. S. (1955) 'Comparative agricultural potentials in the world's regions', *Econ. Geogr.*, **31**.

VOPPEL, G. (1961) *Passiv- und Aktivräume und verwandte Begriffe der Raumforschung im Lichte wirtschaftsgeographischer Betrachtungsweise*, Bad Godesberg.

WAGEMANN, E. (1943) 'Das Alternationsgesetz wachsender Bevölkerungsdichte', *Forsch. und Fortschritte*, 113–15.

WAIBEL, L. (1927) 'Die Sierra Madre de Chiapas', in *Abhandlungen Deutscher Geographentag*, Karlsruhe, 87–98.

WAIBEL, L. (1930) 'Die wirtschaftsgeographische Gliederung Mexicos', *Festschrift f. A. Philippson*, Leipzig, 32–55.

WAIBEL, L. (1933) *Probleme der Landwirtschaftsgeographie*, Breslau.

WAIBEL, L. (1937) *Die Rohstoffgebiete des tropischen Afrika*, Leipzig.

WAIBEL, L. (1939) 'White settlement in Costa Rica', *Geogr. Rev.*, 529–50.

WAIBEL, L. (1948) 'A Teoria de von Thünen sobre a Influência da Distância do Mercado Relativemente a Utilicacao da Terra—Sua Aplicacao a Costa Rica', *Rev. Brasileira de Geografia*, **10**, 3–32.

WAIBEL, L. (1950) 'European colonization in Southern Brazil', *Geogr. Rev.*, **40**, 529–47.

WAIBEL, L. (1955) 'Die europäische Kolonisation Südbrasiliens', *Coll. Geogr. Bonn*, **4**.

WANDER, H. (1965) 'Die Beziehungen zwischen Bevölkerungs- und Wirtschaftsentwicklung, dargestellt am Beispiel Indonesiens', *Kieler Studien*, **70**, 279.

WATTERS, R. F. (1960) 'The nature of shifting cultivation', *Pacific Viewpoint*, **1**, 59–99.

WATTERS, R. F. (1960) 'Some forms of shifting cultivation in the South-West Pacific', *Journ. Trop. Geogr.*, **14**, 35–50.

WATTS, D. (1972) 'Adjustments of Small-Scale Farmers to the Abandonment of of Tropical Estate Agriculture—The Case of Nevis, West Indies', *International Geography* (papers submitted to Int. Geog. Congr., Montreal), 1069–70.

WEBSTER, C. C. and WILSON, P. N. (1966) *Agriculture in the Tropics*, London, Longmans.

WEIGT, E. (1965) 'Entwicklungsländer und ihre Bedarfsstruktur unter dem Gesichtspunkt der Exportmarktforschung', in *Festschrift Leopold Scheidl*, Vienna, 126–39.

WEISCHET, W. (1970) *Chile, seine länderkundliche Individualität und Struktur*, Darmstadt.

WERTH, E. (1954) *Grabstock, Hacke, Pflug: Versuch einer Entstehungsgeschichte des Landbaus,* Ludwigsburg.

WEST, H. W. (1963) 'Reflections upon the problems of land registration in Buganda', in Proc. *East Afr. Inst. Soc. Res. Conf. Kampala*.

WHARTON, C. R. jnr. (1969) *Subsistence Agriculture and Economic Development*, Chicago, Aldine Publishing Co.

WHITE, C. M. N. (1963) 'Factors determining the content of African land tenure systems in Northern Rhodesia', in Biebuyck (1963), 364–73.

WHITE, H. P. and VARLEY, W. J. (1958) *The Geography of Ghana*, London, Longmans.

WHITTLESEY, D. (1937) 'Shifting cultivation', *Econ. Geogr.*, **13**, 35–52.

WHYTE, R. O. (1967) *Milk Production in Developing Countries*, New York, Praeger; London, Faber.

WILBRANDT, H. (1966) 'Welternährung als ökonomisches und soziales Problem', in *Archiv der DLG*, **37**, *Vorträge der deutschen Landwirtschaftsgesellschaft*, Frankfurt/M. 1966, 55–97.

DE WILDE, J. C. (1967) *Agricultural Development in Tropical Africa*, Baltimore.

WILHELMY, H. (1940) 'Probleme der Urwaldkolonisation in Südamerika', *Zeitschr. d. Gesellschaft f. Erdkunde*, Berlin, 303–14.

WILHELMY, H. (1949) *Siedlung im Südamerikanischen Urwald*, Hamburg.

WILHELMY, H. (1954) 'Die Weidewirtschaft im heissen Tiefland Nordkolumbiens', in *Geogr. Rundschau*, **6**, 41–54.

WILHELMY, H. (1963) 'Agrarsoziales Gefüge und landwirtschaftliche Betriebstypen der La Plata Länder', *Mitt. Osterreich. Geogr. Ges.*, **105**, 53–64.

WILHELMY, H. (1966) 'Tropische Transhumance', *Heidelberger Studien z. Kulturgeographie*, Wiesbaden, **15**, 198–207.

WILHELMY, H. (1970) 'Amazonien als Lebens- und Wirtschaftsraum', *Deutsche Geogr. Forschung in der Welt von heute*, Kiel, 69–84.

WILLIAMSON, A. V. (1930) 'Indigenous irrigation works in peninsular India', *Geogr. Rev.*, **20**, 21.

WILLIAMSON, G. and PAYNE, W. J. A. (1959; 2nd edn. 1965) *An Introduction to Animal Husbandry in the Tropics*, London, Longmans.

WILMET, J. (1958) 'Essai d'une Ecologie humaine au Territoire de Luiza (Kasai–Congo Belge)', in *Soc. Belge d'Etud. Geogr.*, **27**, 307–63.

WILMET, J. (1963a) 'Contributions récentes à la connaisance de l'Agriculture itinérante en Afrique Occidentale et Centrale', *Soc. Belge d'Etud. Geogr.*, **32**, 51–63.

WILMET, J. (1963) 'La réparation de la population dans la dépression des rivières, Mufuvya et Lufira (Haut-Katanga)', *Acad. Roy. Sci. d'Outre-Mer*, 247.

WILMET, J. (1963) 'Systèmes agraires et techniques agricoles au Katanga', *Acad. Roy. Sci. d'Outre-Mer*, Brussels, 1963.

WIRTH, E. (1956) 'Der heutige Irak als Beispiel orientalischen Wirtschaftsgeistes', *Die Erde*, **8**, 30–50.

WIRTH, E. (1962) 'Agrargeographie des Irak', in *Hamburger Geograph. Studien*, **13**.

WIRTH, E. (1965a) 'Zur Sozialgeographie der Religionsgemeinschaften im Orient', in *Erdkunde*, **19**, 265–84.

WIRTH, E. (1965b) 'Religionsgeographische Probleme am Beispiel der syrisch-libanesischen Levante', in *Abhandlungen Deutscher Geographentag Bochum*, 360–70.

WIRTH, E. (1966) 'Über die Bedeutung von Geographie und Landeskenntnis bei der Vorbereitung wirtschaftlicher Entscheidungen und bei langfristigen Planungen in Entwicklungsländern', in *Nürnberger Wirtschafts- und Sozialgeogr. Arbeiten*, Nürnberg, **5**, 77–83.

WIRTH, E. (1969) 'Das Problem der Nomaden im heutigen Orient', *Geogr. Rundschau*, 1969, 41–51.

WISSMANN, H. VON (1948) 'Pflanzenklimatische Grenzen der warmen Tropen', *Erdkunde*, **2**, 81–92.

WISSMANN, H. VON (1966) 'Klimagebiete der Erde Kartenbeilage', in *Blüthgen*, 1966.

WOERMANN, E. (1959) 'Landwirtschaftliche Betriebs systeme', *Handwörterbuch der Sozialwissenschaft*, Stuttgart, Tübingen, Göttingen, **6**, 476–92.

WORLD BANK (1972) *Agriculture Sector Working Paper*, Washington.

WRIGHT, R. L. (1972) 'Some perspectives in environmental research for agricultural land-use planning in developing countries', *Geoforum*, **10**, 15–33.

WRIGLEY, G. (1971) *Tropical Agriculture: the development of production*, London, Faber.

ZIMMERMAN, L. J. (1965) *Poor Lands, Rich Lands: the widening gap*, New York, Random House.

Index